Leadership in Chaordic Organizations

Beverly G. McCarter
Brian E. White

CRC Press
Taylor & Francis Group
Boca Raton London New York

CRC Press is an imprint of the
Taylor & Francis Group, an **informa** business
AN AUERBACH BOOK

Complex and Enterprise Systems Engineering Series

First published 2013 by Auerbach

Published 2019 by CRC Press
Taylor & Francis Group
6000 Broken Sound Parkway NW, Suite 300
Boca Raton, FL 33487-2742

© 2013 by Taylor & Francis Group, LLC
CRC Press is an imprint of the Taylor & Francis Group, an informa business

No claim to original U.S. Government works

ISBN-13: 978-1-4200-7417-8 (hbk)

Visit the Taylor & Francis Web site at
http://www.taylorandfrancis.com

and the CRC Press Web site at
http://www.crcpress.com

To Alex and Carol
Alexander Conner McCarter Brueggeman (aka, Alundiel)
October 30, 1989—February 10, 2010
non omnis moriar

—Quintus Horatius Flaccus

Carol Kay Carlson White
December 16, 1939—September 15, 2009
Oh, where are you Carol, my entwined soul?
Your memory trumps
Grief, loneliness, fear of death.
Together, what aspiring lives we shared!
Love—youthful, growing, enduring
Children—centering, nurturing, sustaining
And grandchildren remembering your
Playful, loving, generous, encouraging nature.
You live on in everyone's heartfelt thoughts
—Family, kin, friends, colleagues—
All touched by your gentle spirit.
A brand of eternal life I hope to taste forever.

—Brian Emery White

Contents

Foreword

Why do engineering development projects fail? The large body of literature on systems engineering management, program management, project management, and product management describes systems development in terms of people, processes, procedures, methods, and tools, all precisely laid out using a reductionist, deterministic divide-and-conquer approach as if all one had to do was "follow the yellow brick road." These books acknowledge the complexity of the systems being developed and the desire for systems to meet functional, suitability, and quality requirements.

Leadership in Chaordic Organizations addresses the fundamental issue of the human condition in systems development that transforms what is assumed to be a deterministic problem not into a stochastic problem, but rather a nondeterministic problem. The authors define complexity and its impact on the organization, lay out the problem and provide workable solutions, describe the human condition, show how to build trust, give us a new vision of high-performance teams, and illuminate theory with applications. They finish with a discussion of wicked problems and the potential of multi-user virtual environments. The result is a transformative vision of the future of engineering development teams and a timely contribution to the systems engineering leadership initiative that is now underway to more effectively influence programs for more successful outcomes than heretofore.

I have several decades of professional systems engineering experience spanning the Bell System (a regulated telecommunications monopoly) and transitioning to a competitive telecommunications marketplace, the aerospace/defense domain, and graduate systems engineering education. Gore Vidal's novel, *Lincoln* (Knopf Doubleday, 2005), contains this wonderful passage: "Mr. Seward, the inability of men to grasp an obvious truth is a constant in political life. I seem to spend most of my time explaining what should be obvious to all," speaking to one of the members of what Doris Kearns Goodwin describes as a *Team of Rivals* (Simon & Schuster, 2005). I offer this observation of the protagonist in Vidal's novel from the political domain to the engineering domain, looking back across the decades as an engineering practitioner, engineering manager, engineering consultant, and graduate engineering educator, as well as my volunteer service to the International Council on Systems Engineering (INCOSE). Patience, persistence, and the awareness that we all have value are virtues to be nurtured, thereby enriching our lives as we succeed in our enterprises.

Some years back, I silently observed a multi-hour meeting of engineering functional managers on an integrated product team (IPT) for a development program. The IPT leader, a functional manager at the same management level as the other functional managers on the team, tried to impose his vision of the system concept for integration of commercial off-the-shelf technology, custom hardware, and software. Over the course of the meeting, the managers verbally disagreed with each other, not realizing they were in fundamental agreement. The decibel level in the room, aptly designated the "war room," increased as disagreements became personal and expletives flew in all directions. When there was nothing left to say, stillness came over the meeting. I rose from the workstation where I was sitting and wrote with black marker at the top of a whiteboard: "Comprehension and enlightenment are asynchronous.

Not everybody gets it at the same time." A day or so later I appended: "… and some never get it at all." My comments remained for over a year until we moved to a different facility. Shortly thereafter, the IPT systems engineering manager (who was not the IPT lead) moved out of the IPT to be the systems engineering manager over several IPTs, and I (a consultant) was named the IPT systems engineering lead, with cost account manager responsibilities to go with the position. I experienced the customary 80/20 distribution of my time that seems to have been a constant throughout my professional career: 80 percent of my time working people issues one-on-one and 20 percent addressing technology issues. The IPT did muddle through insurrections against the IPT lead (me), the usual technology travails, and the program manager adding *value* to our lives when things did not go precisely according to plan. As a team, we successfully developed a product that was on time, under budget, met its requirements, and is still in operation today. My considerable personal attention to people issues was critical to this desirable outcome.

Any attempt to *cookie cutter* human beings into deterministic automatons is doomed to fail. For that we await *the rise of the machines.*

William (Bill) D. Miller
Executive Principal Analyst, Innovative Decisions, Inc.,
Vienna, Virginia
Deputy Technical Director, International Council on Systems
Engineering

Foreword

For this book, Beverly Gay McCarter and Brian E. White have
gathered, organized, and interpreted an immense amount
of information. They bring together the thinking from chaos
and complexity theory, psychology, neuro-linguistic program-
ming, and neuroscience, helping readers better understand the
nature and challenges of leadership and improving their effectiveness in a dynamic, rapidly changing, and interconnected
world.

Organizations are collections of self-organizing people talking, interacting, and bumping into each other as they try to
make a living and assist the organization in fulfilling the needs
of its stakeholders. The people are autonomous to various
degrees, behaving in complex ways. They are a whole system
of complex, dynamical interactions. They are hungry for new
information about what is happening within and outside the
organization.

Everything begins with and happens through people
in all their interactions and conversations. Information is
shared and new insights and potential opportunities are
developed. When there is a high level of trust and interdependence, the people can function at their best, focusing
their energy and creativity toward meeting and exceeding
the stakeholders' needs. The people in these kinds of organizations are the most productive, efficient, effective, and
humane. Their conversations about the critical issues facing them are held in an environment of openness as they

explore together the best solutions to their challenges. New information is constantly created during these interactions that improves their decision making and competitive advantage. The people in these organizations move with nimbleness and decisiveness to take advantage of the new potential created as they work together.

When the people in the organization have co-created a shared sense of clarity, focus, and direction, they can put all their energy and creativity into achieving excellence in all they do. Most of the costs related to adverse events, controversial incidents, dysfunctional behaviors, and resistance to change disappear. Everyone is freed to create a brighter, more sustainable future for all the stakeholders, thus generating new and expanding potential.

The above representation, the essence of business leadership in any organization of people, can be captured and modeled in the Process Enneagram© provided here:

The Leadership Process©: The Four Simple Parts to Leadership Excellence

1. Get clear, focused, and determined.
   a. Co-create the living strategic plan using the Process Enneagram.
   b. Keep it posted, talked about, and used.
2. Build trust and interdependence.
   a. Develop the shared, co-created principles and standards of behavior that are needed to achieve excellence in all aspects of the business.
   b. Live by them in doing the work on the issues that need to be addressed to improve performance.
   c. Hold each other accountable.
   d. Let everyone know you deeply care about the core objectives; e.g., safety and everyone going home healthy and in one piece.
   e. Make the organization's work open and visible for everyone to see and model this in your own work.

    f. Help trust and interdependence emerge as people learn to work together this way.

3. Talk with everyone, share information openly, and listen to each other.

    a. Walk around, truly listening, talking, and exchanging ideas, every day.

    b. Observe how mutual trust builds as people get to know you better, see you being honest and keeping your word, trying to improve yourself, and admitting to mistakes when they are made.

    c. Learn how the resulting patterns of behaviors indicate areas of strength and weakness that need to be addressed. Perhaps there are bad habits where more training is needed, or confusing mixed messages, or deeper systems problems that need to be straightened out.

    d. When people become aware of and understand what is happening, apply focused attention constructively.

4. Take appropriate action. Do Steps 1, 2, and 3 all at the same time. They are all interconnected and interacting all the time. Do them over and over again. It requires four Cs:

    a. *Courage* to hold each other accountable, to have the difficult conversations, to make decisions, and to act.

    b. *Care* to do everything as well as you can.

    c. *Concern* for the impacts of all the changes on all the stakeholders.

    d. *Commitment* to stay the course in both the good and difficult times, day after day, month after month, and year after year.

This leadership process is a method for developing a culture that supports the people in co-creating their future, helping to release the creative energy and commitment needed to achieve excellence. Missteps and dysfunctional behaviors such as resistance to change are big wastes of money, time, and energy.

This leadership process is also a whole and complete cycle. It is a living process for growth and learning. After each complete cycle, a new cycle begins again, moving the organization to even higher levels of excellence. It is about engaging with people in organizations in authentic and purposeful ways and bringing out the best in people to achieve business excellence. Leadership is about creating the conditions that enable people to become their best. Leaders can create a virtual, porous container that I call the Bowl to hold the system and provide the focus for their work to succeed.

The best organizations in which I have worked were ones where there was a sense of wholeness in the purposeful way people behaved and worked, greater profitability and return on investment, and improved effects on the environment. These were organizations that people cared about and that built their sustainable future.

Leadership in Chaordic Organizations provides the scientific support for this way of thinking and leading. This book contains critical information to help leaders *understand* and use that understanding to *act* in helping their organizations become more effective, more efficient, and more sustainable. These ideas will also further help leaders dispersed across the Internet in our growing virtual environment to work more effectively in creating a more sustainable world.

Richard (Dick) N. Knowles
Principal and Organizational Anthropologist,
Richard N. Knowles & Associates, Inc., Youngstown, New York

Preface

This book is intended to influence the mindsets of members of almost any organization, particularly leaders. We hope to stimulate a better understanding of complex system behaviors and suggest ways of improving operations through the application of complex systems engineering principles and psychological counseling techniques.

The book contains six chapters. Chapter 1 is about complexity, complex systems, and complex systems engineering, and their potential impacts on organizations. Chapter 2 discusses what it means to be human in an organizational context—how we perceive our environment and those in it, how we are perceived by others, and how we think about and react to workplace situations. Chapter 3 is mainly about trust—what it is and how it is attained (and sometimes lost). Chapter 4 reviews and focuses on group dynamics in an organizational context. Chapter 5 brings the preceding theories together and outlines practical processes for achieving greater organizational effectiveness. Chapter 6 expands on the use of virtual worlds to help in these efforts and references other sources where readers may learn more.

We hope this treatment is sufficient to cultivate and nurture readers' ideas in organizational development. We welcome comments and interactive involvement that would help further this work.

Acknowledgments

As a special acknowledgment, I would like to thank my co-author, Brian, for his heroic patience and efforts to push the book to completion despite the personal tragedies we both experienced in our lives during the writing of the book. The book would never have been finished without his diligent stewardship.

B. G. M.

Many thanks to the MITRE Corporation's Paul Garvey, a co-editor of our Taylor & Francis book series, *Complex and Enterprise Systems Engineering,* for his encouragement and support during the creation of this book idea and its development and completion. Several colleagues at the MITRE Corporation reviewed various versions of the book chapters. I gratefully thank Duane Hybertson (Chapter 1), Karen Detweiler and Dolly Greenwood (Chapter 3), and Gene Pierce (Chapter 4) for their extensive comments, which led to greatly improving the texts.

Also, we greatly appreciate the guidance and patience of Taylor & Francis' Rich O'Hanley, who made this book possible, after considerable delays in its preparation due to personal setbacks in the authors' lives as well as other significant professional commitments.

B. E. W.

Authors

Beverly Gay McCarter is an award-winning architect of immersive virtual environments whose company, Human Mosaic Systems (www.HumanMosiacSystems.com), is located in Cary, North Carolina (Research Triangle Park). She received her MS degree in counseling psychology and human systems from Florida State University and her MFA in studio art from the Memphis College of Art. She is also certified in the areas of facilitating self-organizing systems for complex environments (the Center for Self-Organizing Leadership) and the design and architecture of virtual worlds (University of Washington).

While working as an independent contractor at the National Defense University in Washington, DC, she instructed faculty and Pentagon leadership in the navigation and use of virtual worlds, in addition to coordinating and executing the efforts of the Federal Consortium for Virtual Worlds to bring the federal government into virtual worlds. McCarter is an architect/designer of 3D immersive virtual environments focusing on the psychology of the avatar and virtual worlds, the inherent complex dynamics involved, as well as the impact of the aesthetics of 3D immersive environments on complex human systems. She has advised the Pentagon virtual simulation teams, a solution provider for Linden Labs' Second Life virtual world, and currently works in a variety of other immersive virtual platforms. As a facilitator, McCarter has worked with the Smithsonian Institution in Washington, DC, among others, facilitating group discussions untangling the "wicked problems" of social interactions that overwhelm today's organizational structures.

McCarter is an award-winning artist, focusing on human dynamics and the inherent effects of complexity and the edge-of-chaos on human consciousness. Her work emphasizes the ability to "see" multiple perspectives and the complexity of who we are, and our relationships with others.

Through the integration of these areas of expertise, McCarter designs powerful immersive environments that enable self-organization and collaboration as well as a deeper understanding about one's own perspectives or views of reality and those of others. She architects environments that understand the underlying complex human dynamics and help facilitate the building of relationships, bonding, and greater understanding, all of which help individuals collaborate as they make decisions in complex, ever-changing environments.

A sample of videos that expand more on McCarter's work can be found at the following link: http://www.humanmosaic-systems.com/page1.php.

Relevant publications and presentations by McCarter include

- Lowell F. Christy, Jr., and Beverly G. McCarter. "Why Whole Systems Thinking Is Moving Beyond Engineering Backwaters." George Washington University. Research Program in Social and Organizational Learning. Washington, DC. January 29, 2008.
- B. G. McCarter, "Federal Virtual World Challenge: Description, Results and Discussion," Federal Consortium for Virtual Worlds, Washington, DC. May 16, 2012.
- B. G. McCarter, "Immersive Intelligence and 3D Data Visualization," Federal Consortium for Virtual Worlds, Washington, DC. May 16, 2012.
- B. G. McCarter, "Narrative Structures, Wicked Problems, and MUVEs," East Coast Gaming Conference, Raleigh, NC. April 25–26, 2012.

- B. G. McCarter, "Federal Virtual World Challenge: Winner Demonstrations," GameTech Users' Conference, Orlando, FL. March 28–30, 2012.
- B. G. McCarter. "Designing and Facilitating for Living Human Systems: Wicked Problems and MUVEs." Second Life Community Convention. Oakland, CA. August 14, 2011.
- B. G. McCarter. "Walking across Government Silos: The Power of Virtual Worlds." FISSEA. Washington, DC. March 25, 2010.
- B. G. McCarter. "Center for Naval Intelligence: Virtual World Overview." Center for Naval Intelligence. Virginia Beach, VA. March 23, 2010.
- B. G. McCarter. "Reaching across Organizations with Virtual Worlds." Social Media for Government. Washington, DC. March 12, 2010.
- B. G. McCarter. "Virtual Simulations and Decision Making." Innovative Decisions, Inc. Vienna, VA. May 2009.
- B. G. McCarter. "Complex Human Systems: Paradigm Shifts through Immersive Virtual Environments." United Nations Climate Change Conference (COP15). Virtual World: Second Life. December 9, 2009.
- B. G. McCarter. "Complex Human Systems and Immersive Virtual Environments." 2nd Louisiana Conference on Virtual Worlds and Higher Learning. Virtual World: Second Life. November 2009.
- B. G. McCarter. "Social Aspects of Complexity: 21st Century Dynamics." Applied Systems Thinking Institute. Washington, DC. September 2009.
- B. G. McCarter. "Social Aspects of Complexity: Self-Organizational Dynamics." 2008 Understanding Complex Systems Symposium. University of Illinois at Urbana-Champaign. May 2008.
- B. G. McCarter. "Complexity and Managing Change in Organizations." George Washington

University. Research Program in Social and Organizational Learning. Washington, DC. March 2008.

- Beverly G. McCarter. "Social Aspects of Complexity." MITRE (Boston, MA, McLean, VA, and Chantilly, VA). International Council on Systems Engineering New England Chapter (Boston, MA). March 12, 2008.
- B. G. McCarter and B. E. White. "Emergence of SoS, Socio-Cognitive Aspects." In *System of Systems Engineering—Principles and Applications*, edited by M. Jamshidi. Boca Raton: CRC Press. 2009.
- B. G. McCarter and B. E. White. "Collaboration/ Cooperation in Sharing and Utilizing Net-Centric Information." 5th Annual Conference on Systems Engineering Research. March 2007.
- William D. Miller, Gay McCarter, and Craig O. Hayenga. "Modeling Organizational Dynamics." IEEE International Conference on Systems Engineering. SMC. Los Angeles, CA. April 2006.

McCarter's art-related publications include

- "A Mosaic for Jackson." *Southern Living Magazine.* February 2002.
- Featured artist in *Mosaic Techniques & Traditions: Projects and Designs from Around the World.* Compiled by Sonia King. New York: Sterling Publications, September 2002.
- *Mosaic Matters International Online Magazine.* November 2001.
- *Tennessee Arts Commission Magazine.* December 2000.
- *Jackson Chamber of Commerce Magazine.* December 2000.
- Public Art Commission. *Jackson and the Arts: An Outdoor Mosaic Wall Mural.* Ned McWherter West Tennessee Cultural Arts Center. Jackson, TN. 2000.

■ Book cover for *Social Welfare: A World View*. By Katherine Van Wormer. Chicago: Nelson-Hall, 1997.

Brian E. White received PhD and MS degrees in computer sciences from the University of Wisconsin and MS and BS degrees in electrical engineering from MIT. He served in the U.S. Air Force and for eight years was at the MIT Lincoln Laboratory. Dr. White was a principal engineering manager at Signatron, Inc., for five years. In his twenty-eight years at the MITRE Corporation, he held a variety of senior professional staff and project/resource management positions. He was director of MITRE's Systems Engineering Process Office from 2003 to 2009. White left MITRE in July 2010 to establish a consulting service, CAU←SES ("Complexity Are Us" ← Systems Engineering Strategies).

White strives to help people change or improve their mindsets by communicating complex systems behaviors and complex systems engineering research knowledge, precepts, and principles where people (particularly stakeholders) are considered part of a system, system of systems, enterprise, or complex system. He accomplishes this by drawing on his twenty-eight years of experience working with corporate executives, middle managers, project leaders, technical staff, and academic colleagues in areas of complexity theory, complex systems, and complex systems engineering.

A summary of Dr. White's professional activities in recent years follows:

2010–2011: International Council on Systems Engineering (INCOSE)
  – Presented tutorial on Principles of Complex Systems Engineering at INCOSE Symposium 2011 (June 2011), INCOSE Washington (DC) Metropolitan Area Chapter (August 2011)

- Rejoined INCOSE as senior member in July 2010. For INCOSE Symposium 2010, was invited to be on two panels (architecture and cyber-security). Presented his 2010 IEEE International Systems Conference paper in the IEEE session and chaired the above session and a complex systems session.

2006–present: Co-editor of a MITRE book series, *Complex and Enterprise Systems Engineering*, with Taylor & Francis.

- Co-author of chapter: "Emergence of SoS, Socio-Cognitive Aspects." In *System of Systems Engineering— Principles and Applications*. Boca Raton, FL: CRC Press, 2009.
- Co-editor and chapter author: "Enterprise Opportunity and Risk." In *Enterprise Systems Engineering: Advances in the Theory and Practice*. Boca Raton, FL: CRC Press, 2011.
- Co-editor of book: *Case Studies in System of Systems, Enterprises, and Complex Systems Engineering*. Expected publication date: late 2013.

2005–present: Authored and presented more than twenty conference papers in complex systems engineering. Selected citations:

- B. E. White. "Managing Uncertainty in Dating and Other Complex Systems." Conference on Systems Engineering Research (CSER). Redondo Beach, CA. April 15–16, 2011.
- B. E. White and P. N. Jean. "Case Study in System of Systems Engineering: NASA's Advanced Communications Technology Satellite." 6th IEEE International Conference on System of Systems Engineering (SoSE 2011). Albuquerque, NM. June 27–30, 2011.

- B. E. White. "Let's Do Better in Limiting Material Growth to Conserve Our Earth's Resources." Conference on Systems Engineering Research (CSER). St Louis, MO. March 19–22, 2012.
- B. E. White et al. "Application of Case Studies to Engineering Management and Systems Engineering Education." Annual Conference, American Society of Engineering Education. San Antonio, TX. June 10–13, 2012.
- B. E. White. "Systems Engineering Decision Making May Be More Emotional than Rational." Annual INCOSE International Symposium. Rome, Italy. July 9–12, 2012.

2003–present: Further developed and augmented MITRE's "Regimen for Complex Systems Engineering" to create what is now called the Complex Adaptive Systems Engineering (CASE) methodology.

- Developed case studies showing the degree to which CASE principles are followed in important government programs such as maritime domain awareness and Census Bureau information processing.

Chapter 1

Definition of *Complexity* and Its Impact on Organizations

B. E. White

We use the term *complex* to describe people, their perceptions, motivations, mindsets, personalities, and so forth, but primarily their behaviors. People's actions, as expressed by verbal, written, and body language and physical activities, are what impact others directly. Organizations include people and thus are also complex. However, as Covey states, "In a very real sense there is no such thing as organizational behavior. There is only *individual* behavior collectivized in organizations" (Covey, 2004, p. 102). Typical organizations try to utilize people, processes, and technologies to accomplish specific goals. Often there are "systems" involved. A system can be thought of as a collection of elements (e.g., people, processes, and technologies) that is intended to achieve a purpose that is greater than any subset of the elements. When systems include people, they are also complex.

I credit Yaneer Bar-Yam for introducing me to the world of complexity theory, as a discipline, through a four-day short course presented at the MITRE Corporation in June 2003. Bar-Yam's work (2004) is recommended background reading on the subject of how complex systems concepts can help in solving complex problems in a decentralized world. In another work, Bar-Yam (2002) provides a broad introduction to complexity, from an individual human perspective, through organizational hierarchies and networks to human civilizations. Douglas O. Norman and Michael L. Kuras are two MITRE colleagues who interacted with Bar-Yam and contributed their own added value to the topics of complexity, complex systems, and especially complex systems engineering (CSE) (Norman and Kuras 2004). Norman introduced a regimen for CSE and Kuras elaborated on it.[*]

No one is in total control of another person. It follows that no one is in total control of an organization or any system that includes people. Thus, organizational leaders can hope only to influence or leverage others, not control them. Those who try to impose control may achieve some limited success temporarily but fail in the longer term because people ultimately resent being treated like machines. Covey (2004) explains this well and suggests that each of us becomes more aware of detrimental codependencies in organizations. This dynamic occurs when formal leaders try to control or manage their (follower) workers by imposing orders and providing carrot-and-stick rewards and punishments from the top and when people respond by doing only what they are told and absolving themselves of any responsibility for taking independent, creative action. Enlightened leaders recognize this and focus more on harnessing the motivations of their people by creating conditions conducive to progress for the organization through shaping people's self-interest or at least the interest of groups. Covey (2004) suggests that everyone can become a leader through the exercise of choice, his "8th habit," characterized

---

[*] I am grateful for everything I learned about complexity from both colleagues, principally during our 2003–2008 interactions.

by finding one's voice and inspiring others to find theirs. Individuals are more likely to perform* according to how they or their groups are measured and what is in it for them, primarily as groups but also as individuals. I like to think of this leadership role as "engineering" the *environment* of a complex system rather than trying to engineer the systems themselves. Hence, we embrace CSE.

It may seem strange to use the term engineering. However, in common parlance engineering is about trying to create solutions to practical problems using whatever methods might be readily available. This includes methods that come from various sciences that are each focused on understanding nature, in this case mainly the nature of people. Individual philosophy (Boardman and Sauser 2008), psychology, group dynamics, social science, and organizational theory are fundamental to the study of complex systems and the formulation of promising CSE practices.

The hope is that by understanding complexity theory, complex systems, and CSE better, humankind will be better equipped to solve the most important complex problems facing the human race and the world. This is the optimistic view. There are those, however, who take an extremely pessimistic viewpoint that life itself (of all forms, not just humanity) "screws up" the world's works (Bennett 2009) and will continually defeat us in our aspirations to make the world a better place.

In this chapter we delve into the science of complexity a little deeper, relate its principles to systems of people, and then discuss the impact on organizations.

---

* Could it be that the vast majority of workers are content in just wanting a job within almost any organization and keeping their "head down," letting those in authority dictate things, without feeling the need to be creative in changing their environment? If so, this is the kind of cultural inertia that sabotages constructive organizational change and profoundly challenges would be formal leaders.

Definition of Complexity*

Complexity is one of those words that should be often avoided because people interpret it in various ways, from "complicated" or "mind-boggling" to something that seems to have a "life of its own" that we cannot begin to understand. In this conversation, complexity does *not* mean complicated (i.e., something that is so detailed and intricate that it is beyond the typical person to grasp or follow it), but rather something that seems "alive." One is sometimes tempted in the extreme to label observed complex behaviors as "randomness" or "chaos" if there seems to be no viable explanation.

The complexity of complex adaptive systems can usually be relegated to somewhere in the vast region between stasis (equilibrium or complete order) and chaos. Another colleague of ours, Sarah A. Sheard [a fellow of the International Council on Systems Engineering (INCOSE)], has used Figure 1.1 to describe this relationship. Complex systems demonstrate a

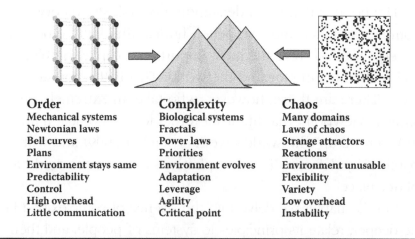

Order	Complexity	Chaos
Mechanical systems	Biological systems	Many domains
Newtonian laws	Fractals	Laws of chaos
Bell curves	Power laws	Strange attractors
Plans	Priorities	Reactions
Environment stays same	Environment evolves	Environment unusable
Predictability	Adaptation	Flexibility
Control	Leverage	Variety
High overhead	Agility	Low overhead
Little communication	Critical point	Instability

Figure 1.1 Complex adaptive systems adapt toward the "edge of chaos." (From Sarah A. Sheard, Principal, 3rd Millennium Systems LLC. 2008. Used with permission.)

* Detailed definitions and explanations for several key terms used in this chapter may be found at the end of the chapter in Appendix A.

dynamic emergence that is constantly changing and evolving, whereas complicated ones do not. Chaordic systems also demonstrate this "sweet spot" between chaos and order, which enables chaordic organizations to thrive in ever-changing complex environments by being flexible and adaptable as they self-organize. (Also refer to Chapter 5 and in particular, Figure 5.1.) Chaos theory (http://en.wikipedia.org/wiki/Chaos_theory) is important but is relatively difficult to understand. It is also beyond the scope of this book because we do not rely much on its phenomena or results.

Most of the time in our pursuit of "purposeful" and effective systems (and organizations; refer to the next section), we will use the word complex to describe the behaviors of independent autonomous agents (models of people) that interact with each other and their environment. Their motivations for their behavior will be partly rational, partly emotional, and often unpredictable or unfathomable. Sometimes neither we nor they will be able (or willing) to explain thoroughly why they do certain things, even after the fact.

This inability to explain is not just due to shortcomings of rational thought. Descartes was wrong (Demasio 1994). Man is both a rational and an emotional being. It takes a fair amount of emotional intelligence (Goleman 1995), as well as clear analytical thinking, to successfully examine and deduce the underlying causes of our actions. This is especially difficult when we are part of the complex system, as is usually the case, rather than merely an outside observer, which is difficult enough. We have to be aware of ourselves in addition to understanding others in order to gain insights into complex human systems.

The mind, an infinitely complex emergent property of the human brain, is a wonderful thing. This phenomenon is so complex that we do not even venture to define its nature in this book. Rather, we merely allude to some intriguing aspects of how the brain–mind combination might work and the potential implications for our human interactions in face-to-face and distributed communication environments.

We cannot predict how people will make decisions or interact, nor what emergence will occur, through computer simulations in general or even through massive multi-player game simulations.* We can, however, gain a deeper under-standing of how complex human systems behave through such simulations. The Zeitgeist, or environmental factors and cultural influences based on current time and space, is limited to the present only and does not reflect changes that occur in any dynamic system over time (Beverly Gay McCarter, per-sonal communication, October 7, 2008). The environment of the future will be different (of course) and we cannot predict how that difference will be manifested.

As my co-author points out in Chapter 2, scientists are making significant strides in developing reasonable inferences between apparent activity in certain regions of the brain and our ability to function in complex ways. For example, Brown and Parsons (2008) suggest that parts of our brain interact to enable us to dance and keep rhythm, and associated physical gestures were the precursors to speech in our ancient past. Knowledge is processed and retrieved largely through images, and mirror neurons (Ramachandran and Oberman 2006; Rizzolatti, Fogassi, and Gallese 2006) enable us to understand the movements of others as well as the meaning of gestures. Many gestures seem to have a universal translation, although others (e.g., hand signals) are a product of local cultures. Such scientific studies provide clues to the origin of verbal language and the evolution of modern communication that can lead to interpersonal trust and greater mutual understanding. Self-awareness and understanding others are necessary to gain insights into complex human systems.

---

* For example, http://www.kurzweilai.net/superstruct-the-world-s-first-massively -multiplayer-forecasting-game

Watson (2008) refers to the ability of sound to affect human consciousness and even put people into a trance. Such two-way mind–body interactions (e.g., the mind influences our ability to move, to dance to sounds, through a predisposition to rhythm and a need to kinetically experience sound) are not yet well understood. In turn, sound can affect the mind, our experiences, and their recall.

Finally, movie-goers feel how the soundtrack enhances the experience by immersing them more completely into the story being depicted visually on the screen. In Chapter 3 we explore various notions of how all our senses, kinetic experience, and computer-assisted mixed reality states can trigger the generation of oxytocin (an endorphin of the brain), which in turn can engender trust. Further research by scientific, psychological, and sociological specialists undoubtedly will uncover more about what and how we think, trust, and communicate.

Organizational leaders should learn about and become mindful of these phenomena so they can create better opportunities and environments for engendering creative thought, interpersonal trust, and effective communication among their constituents. One way to accomplish this is to encourage storytelling relevant to organizational goals during working hours.

On the other hand, these research efforts probably will also lead to possibilities for mind control, obviously something of which we need to be quite wary.

A system of people tends to be continuously changing and evolving. However, there can be short periods of relative stability. The lengths of these long and short intervals depend on the makeup, size, and distribution of the group and its environment.

When trying to engineer a system (involving people) to solve a complex problem, the usual reductionist (http://en.wikipedia.org/wiki/Reductionism) technique of divide and conquer does not work well.* By this we mean the attempt by a manager or governing body outside the system (of people, processes, and technologies) to perform a sort of work breakdown structure: (1) isolate portions of the system from each other; (2) create fairly well-defined interfaces among the subgroups; (3) accomplish work within each subgroup semi-independently (being mindful of the interfaces); and (4) combine the results to re-form the entire system as an integrated whole. On the surface there seems to be no obvious reason why this will not work. But in practice, such an approach is optimum in terms of achieving overall effectiveness only if each portion or subgroup of the system, as well as the total system, is completely isolated; in other words, *closed.*

What is missing is the recognition that the system (as well as each portion or subgroup of the system) usually is *open,* not closed (i.e., not completely isolated from its counterparts or environment), and that in most practical situations of interest, the counterparts and environment are constantly changing, and so are the external forces on and energy flow into the system. Similarly, the system, in whatever configuration, is interacting with this changing environment. Thus, because of the time delay in disassembling and reassembling the system, the result is unlikely to be well matched to the new environment. In such situations, it may be better to create conditions for many more interactions among the people and with their environment. In this way, human ingenuity and the ability to learn and adapt are unleashed, with the potential to create, over time and in an evolutionary way, more robust solutions to the problem.

An appropriate analogy is the way human languages evolve. No one is in control. The language does what it is

---

* An interesting short article espousing the idea that nature cannot be reduced to mathematical laws due to emergence in complex systems points out the veracity of this even at the molecular or atomic level. Mark Buchanan. "Why nature can't be reduced to mathematical laws." *New Scientist.* October 6, 2008. http://www.newscientist.com/article.ns?id=mg20026764.100&print=true

going to do. People may influence it by introducing or (more often) embracing a new word or phrase, or removing or (more often) dropping an old word or phrase. This process of natural evolution is one of many to study and mimic in CSE work.

Thus, we say that the greater degree of instability in the environment of a system, the more likely that reductionism, and other traditional or conventional systems engineering techniques, will be ineffective in solving the problem. Again, complexity is found in the realm between stasis and chaos where any points of equilibrium are generally or relatively unstable. Furthermore, an individual immersed in a complex environment can have only a limited view and understanding of the complex environment. If one is observing a complex system from the outside, it seems reasonable that there is a greater chance of understanding what is going on, but not necessarily even then. However, it is rare to be outside the system.

It may help the definition of complexity within this context to describe typical complex systems behaviors and some principles to employ when attempting to perform CSE. The following examples were drawn from White (2011b).

Complex System Behaviors

Surprising Emergence

There are many definitions of (positive or negative) emergence (McCarter and White 2009). We favor the definition where emergence is unexpected. Moreover, the most interesting forms are surprises not easily explained; they are worthy of additional effort to further understand complexity.

Evolves on Its Own as a Whole

There seems to be a mysterious intelligence. The system does whatever it pleases whether we try to intervene or not.

The elements interact in ways that make the overall behavior unpredictable, sometimes even when the system is deterministic.* This uncertainty makes complex systems fascinating and challenging.

Acts Robustly

A healthy complex system can survive harsh environments, having already adapted to stress. A *population* of diverse elements survives, although individual members may not without the currently required gene pool (Nicholls 2011). Complex systems may not be efficient, but they can be very effective. When trying to design or upgrade human-made complex systems, foreknowledge of future use can be rather murky. Not only do requirements keep changing, but when fielding, the system may have already outlived its usefulness. Devote considerable attention to flexibility and adaptability of capabilities.

Thrives on Diversity

Many complex systems have an innate beauty manifested by intricate and multifarious interrelationships.

Many Factors at Play

In a team effort, one seeks shared ideas and preferences and becomes acutely aware of discord. Identifying and building common ground is necessary for joint progress.

---

* Even when the next state of a complex system is completely determined by the present state and its input(s), a distant future state often cannot be known in advance. This phenomenon is particularly true when there is great sensitivity to initial conditions, as in the "butterfly effect" (discussed later in this section).

Stimulates Different Perspectives

New ideas continually present themselves, build collectively, and are interrelated, providing greater nuanced understanding and tolerance for uncertainty. Armed with these insights, one can more easily continue to nudge the system in the right direction.

Ever Changing

A complex system has no *status quo*. Recognize the certainty of change and establish a process for managing uncertainty when unexpected events arise. One cannot suggest exact fixes in advance, but this planning lessens the resistance to dealing with significant changes. Just because things seem to be going smoothly and something bad has not yet happened does not imply that it will not. One learns the most from counterexamples to what are (wrongly) considered truisms (Taleb 2007).

Informs the Observer

Dispassionately observing a complex system can illuminate potentially viable interventions. The message can be clearer if one is external because it is more difficult to see the forest for the trees if one is inside. *Systematically learn from experience.* Occasionally standing back and contemplating encounters may clue you in on how to behave differently. Systematic learning from cut-and-try (multiple trial) methods mitigates uncertainty.

Performs Openly

Although the effects of interactions among sub-systems and with the system's environment may be neither obvious nor explainable, the collective behavior is not hidden. Nevertheless, it may still be difficult to interpret the specific causes of the effects observed.

Internal and External Relationships Are Key

Try to ascertain which interactions are more responsible for causing changes. This may entail experimentation to develop fact-based actions, but do not state something as fact if the phenomenon is too confusing.

Self-Organized

Self-organized behavior is typical of complex systems. As in physics, each pairwise interaction may change something associated with either or both elements and/or their affinity for each other. Collectively, these changes create dynamic realignments and patterns that affect system behavior. In a purposeful (self-)organization of humans, solutions to problems not otherwise possible may emerge.

Sensitive to Small Effects

Most people have heard something like, "If a butterfly flaps its wings in Japan, a hurricane may develop in the Atlantic Ocean."[*] This profound effect is not unusual in complex systems; one cannot detail the chain reaction responsible. Also, even slight changes of initial conditions in a chaotic complex system can lead to very different results (as previously discussed). On the other hand, "purposeful" systems are *insensitive* to initial conditions; they move toward a goal regardless of where they started.

Exhibits Tight and Loose Couplings

Complex systems interactions vary dynamically. Pairwise interactions with higher frequency, higher intensity, and/or closer

---

[*] Prusia Buscell. "Surprises Emerge When Little Explodes Into Big," Plexus Associate Thursday Post 8-14-08. Those Big Little Things. http://www.plexusinstitute.org

proximity are said to be tight; those with lower frequency, lower intensity, and/or greater distance are said to be loose. It is natural to group the interactions within each of these two categories for the purposes of understanding, architecture, and/or design. However, this can still be misleading.

Complex Systems Engineering Principles

Bring Humility

This principle has been attacked as unprofessional. What do you think? In conventional systems engineering situations, someone who acts unassertively gets steamrolled by those who tout their solution to a problem with arrogance and confidence. When such dictated approaches do not work, people become less enamored with simple fixes. Thereafter, improvements are pursued more thoughtfully and exercised with greater caution. One must watch what happens and be prepared to try something else. However, one is rarely sure just how long to wait before making the next intervening decision.

Follow Holism

Concentrate on the complex system taken as a whole. For example, endeavor to understand how the behavior of the entire system explains the roles of it components (e.g., subsystems). This may lead to insights about emergent properties that cannot be predicted by the antithesis of holism, namely, reductionism (http://en.wikipedia.org/wiki/Holism, http://en.wikipedia.org/wiki/Reductionism). One cannot use reductionism to accomplish goals. By the time one subdivides the problem, works on optimizing each resulting sub-system, and reassembles the parts, the complex system and its environment have moved on, and little will perform as desired. This is a fundamental problem with government system acquisitions that take many years to accomplish. For example, many overly ambitious weapons programs have been cancelled after billions of dollars have been expended (White 2008).

Achieve Balance

Optimization may be impossible in a mathematical sense. Optimizing sub-systems can detract from the potential efficacy of the whole. Instead, try to balance various sub-system thrusts. In an automobile enterprise, if manufacturing and sales are each rewarded for the most cars, either more cars will be produced than can be sold or so many cars will be sold that manufacturing cannot keep up. Reward collaborations that keep manufacturing and sales abreast while increasing both production and sales.

Utilize Trans-Disciplines

Most engineers think of systems engineering as multi-disciplinary, with the fields of sensing, information processing and computing, communicating, networking, and the hard sciences of physics and mathematics coming together. In CSE, people are considered part of the system. People are difficult, if not impossible, to model or control. Hence, "trans-disciplines," namely, the soft sciences, such as philosophy (Boardman and Sauser 2008), psychology, sociology, organizational change theory, economics, and politics, should be considered.

Embrace POET (Political, Operational, Economic, and Technical) Aspects

Let's face it. In the world's most pressing problems, politics and economics play critical roles, in addition to operational procedures and technical means. CSE must deal with all four POET aspects or results will be unsatisfactory. One may devise a great technical solution that could improve operations, but this will not go anywhere without (political) acceptance from stakeholders. Understand your stakeholders' values to establish win–win scenarios. Someone also needs to agree to pay (economically) for the improvements.

Nurture Discussions

First, realize that every person sees things differently (McCarter and White 2009). No one has an exclusive grasp of the truth about complex systems. Better solutions are attained through leveraging a large group's cognitive diversity than by a panel of experts (Page 2007). The wisdom of the crowd only works when the crowd members do not know each other's opinion (i.e., there is independence of thought) (refer to group dynamics, Chapter 4). Diversity of perspectives enhances creative problem solving but requires a good group facilitator to ensure that the disparate voices and opinions are heard and shared. In systems engineering we spend too much time arguing over definitions instead of seeking to understand how we use words. Only after this mutual understanding is attained can a group make real progress.

Pursue Opportunities

The great emphasis placed on identifying and mitigating risks is often at the expense of pursuing opportunities. In a complex system (such as a military enterprise), the principal risk is *not* pursuing opportunities (White 2011a). Yet a balance must be struck. With many opportunities, the initial expectations of a profitable business relationship can be too high. When visiting a new company, what if grass is taking over the parking lot? Share the impressions of potential prospects in meetings, but do not give away all your advice (especially your intellectual property) for nothing. If successful, no task needing attention is too small; it could lead to other opportunities.

Formulate Heuristics

Knowing when to make what decisions is a formidable challenge in managing uncertainty. Think in terms of pattern recognition and the general rules of behavior that underlie complex human systems. However, be wary of the effects of outlier behavior on the systems as well as your heuristics.* Those in authority must make important decisions, at least occasionally, because it is part of their job. Some excellent work has been accomplished in formulating heuristics to help decision makers (Maier and Rechtin 2009). Nevertheless, heuristics is still a fertile research area. System dynamics was invented by Jay Forrester of MIT by 1968 (Meadows 2008). Here the importance of time delays is paramount. What initially might seem to be positive effects from your system intervention(s) may ultimately prove to be negative, and vice versa. Insist on believable credentials (e.g., a simple heuristic) before engaging. Ask yourself whether the decision makers really get it. Even if there is no burning platform, are decision makers in enough pain because their things are not working to be willing to break their mold and try something different? Heed early signs that involvements are not gelling. Test first impressions while noting whether promises were made good.

Foster Trust

How can one expect to interact with stakeholders productively without mutual trust? Establishing trust is difficult, takes a lot of time, and can be lost immediately if a precipitous event is handled poorly. Once lost, trust is also hard to recover. It is imperative to share information; otherwise, progress will suffer. Do not adopt the usual mantra that information is power, nor follow most organizational

---

* In this chapter, *heuristics* is used in a relatively narrow sense, meaning "rule of thumb." Broader definitions exist; for example, "ways of generating solutions to problems" and "techniques and tools for making improvements" (Page 2007).

cultures of protecting information lest you be punished. Instead, try sharing business information, albeit to a limited extent at first. If this is echoed so you learn more, great! Then share more and more. Be open and honest, freely sharing information.

Create an Interactive Environment

Leaders should not try to drive solutions from the top, for they cannot always know what to do. Instead, continually strive to establish and maintain conditions (e.g., a vision of cooperative interactions and suitable reward structures for doing so) to ensure informed, vigorous, and sustained engagements among the troops. When people play nicely, remarkable improvements are more likely than under autocratic rule. If there is more conflict than competition and collaboration, then either the leader has failed to convey the vision or the incentives are inappropriate and need to be modified. Responsibilities to act and be responsive are important to embrace. Do not follow rules slavishly, but do not fight the system, especially about what you cannot even influence. It may be worthwhile to solicit the inputs of external observers as a check on the organization's direction and/or degree of success.

The following quotations provide support for the power of relationships, interactive environments, self-organization, and diversity:

> When the business was suddenly confronted with a crisis that could destroy it, people came together in a different way and achieved extraordinary things. ... [W]e stopped most of the stupid games we were playing and worked together purposefully. ... The teams then did the work and made many decisions on their own as they went. They often saw better ways to do things and did them. ... Everyone had access to everyone. As we shared the information

about our progress, we just kept learning from our mistakes and successes. [A]ll the systems were full of constant feedback. ... [W]e co-created our future together. People didn't resist the changes they were helping to create. (Knowles 2002, 2, 26)

... diversity and accuracy contribute equally to collective predictive performance, and that a crowd's collective prediction must always be at least as good as the average prediction of a member of the crowd. ... three core claims: (1) Diverse perspectives and tools enable collections of people to find more and better solutions and contribute to overall productivity; (2) Diverse predictive models enable crowds of people to predict values accurately; and (3) Diverse fundamental preferences frustrate the process of making choices. (Page 2007, 10, 25)

Stimulate Self-Organization

A hallmark of complex systems is self-organization among its living elements. This is particularly true in natural complex systems such as beehives, anthills, bird flocks, fish schools, and so forth. Human languages also strongly self-organize based on how people talk.* The powerful human rights demonstrations in the Middle East in early 2011 were self-organized (using social networking).

---

* Chomsky is widely known for espousing (perhaps erroneous) a universal grammar about the singularity of human language, suggesting a first language from which all others arose ... and pointing to common patterns that lead back to the original (http://en.wikipedia.org/wiki/Noam_Chomsky). This contrasts with the Whorfian hypothesis (repeated in Bertalanffy 1968, 222–223), "...that the linguistic patterns themselves determine what the individual perceives in this world and how he thinks about it. Since these patterns vary widely, the modes of thinking and perceiving in groups utilizing different linguistic systems will result in basically different world views" (Fearing 1954).

Seek Simple Elements

Another property of many complex systems in nature is the simplicity of the participating elements. For example, it is amazing what can be accomplished in beehives and anthills within their robust societies when each worker bee or ant seems so limited in capabilities. Imagine what humans might be able to do when they interact in self-organizing ways! One difficulty with intended systems engineering solutions to complex problems is the size and complexity (or maybe just complication) of the individual pieces. Current chaordic (collaborative and competitive) efforts also point to this. It might be better to design down-scale and assemble arrangements of smaller identical units that are good at adapting.

Some complex systems consist of simple elements, but admittedly many complex systems consist of complex elements. The implication of this *seek simple elements* principle is to consider how one's organization can be transformed so that it has fewer groupings of simpler elements.

Enforce Layered Architecture

Layering is applied to increase flexibility or introduce system improvements following changes in environment or implementation technology, for example. What might be better realized in software in one era is better done in hardware in the next, and vice versa. Each layer is confined to a set of closely knit basic functions, grouped in categories of applications, networking, communication links, or physical implementations. The interfaces between layers are kept simple and stable. However, the realization within a given layer can be (more often) adapted to different conditions. As long as the interfaces to that layer remain unchanged, the system still operates effectively.

Even complex organizations might be layered to better effect. Members performing similar functions would be tightly coupled within a given layer for high levels of interaction. Each set of fundamentally different functions would be part of its own layer. The various layers would also interact among themselves but with significantly looser coupling.

Human Behavior

A working knowledge and understanding of human psychology and even physiology (e.g., how the brain works) can help individuals work together more effectively. It is important to realize that each person may see things differently, conceptualizing the world through their own cognitive lens. How often do we observe smart people with good ideas being marginalized because of undesirable personality traits that irritate other group members? How often do we see conflict in groups due to differing views or perspectives? Often, differing or dissenting views are ignored or received in a hostile fashion. As a result, consensus is not achieved and those who feel their opinions were not heard will not actively support the group decisions. Part of the challenge is to nurture enough emotional intelligence so that people attack the ideas and not the person. More mature members of the group may need to help achieve and maintain that culture. On the other hand, some members of the group may be resistant to new ideas because of a common phenomenon called "rigidity of thought." Proponents need to learn to present their ideas in ways that are more likely to succeed (Heath and Heath 2007; Covey 2004, 129–131). To the extent feasible (more difficult in a distributed environment), work groups should receive periodic training in the principles of group dynamics. This increases the likelihood of consistently realizing high-performance teams (Nemiro et al. 2008) as people are continually brought together to solve multifarious problems.

Social science can teach us much about what to watch for in and expect from people interactions. Trust, for example, is a huge issue. Two basic kinds of trust relate to competency and vulnerability. The confidence to rely on someone else's input as being valid hinges on the perceived competency of that individual. The other kind of trust, being free from worry that someone else may sabotage one's work or act in an untrustworthy fashion if given the opportunity, is more serious. With this kind of trust, one may risk becoming vulnerable to explore new avenues toward possible solutions to problems without fear of penalty or reprisal. (Refer to Chapter 3 for more about trust.)

Tacit knowledge (Busch 2008) is another important consideration for improving team performance. Much of what people know inherently and even subconsciously about basic assumptions they bring to bear in trying to solve a problem is neither explicitly acknowledged nor documented. It is a challenge of knowledge management to ferret out such tacit understandings so that they can be examined and modified, if necessary, for further progress.

Appendix B is offered as a "sidebar" on the definitions of complexity and complex systems that emerged from efforts of the Systems Science Enabler Group (SSEG) of INCOSE from 2005 to 2008. SSEG essays on complexity, complex systems, and CSE are published in the January 2008 special issue of *INCOSE Insight*, focused on systems science (INSIGHT 2008). A relatively new website called "Codynamics[SM]—The New Way to Think and Work Together" contains some useful introductory material: http://www.codynamics.net/intro.htm

Impact on Organizations

From a formal leader's perspective, the creation of a purposeful organization is a laudable goal, because this means instilling an idea of vision and mission that can be internalized by the people of the organization in a way that encourages them to assume leadership roles as well. Such a formal leader will encourage interactions within the organization as well as pauses for reflection and learning. Like a complex adaptive system, the group will continually self-organize and unexpected collective behaviors will emerge. It is important to create a culture for and to be open to change, continually learning, adapting, and making decisions for the future. As will be explored in Chapter 4, creating a purposeful organization is more difficult in a distributed environment.

An example of unexpected behavior came to light during a 14 January 2008 technical exchange meeting (TEM) on organizational learning at the MITRE Corporation. An important U.S. government agency that MITRE supports was facing significant organizational change, and MITRE was trying to help facilitate that transformation. Consider, for the moment, two broad classes of civil servants working in that agency: seasoned veterans with significant tenure and relatively inexperienced professionals who had just joined. Which group was more resistant to the pending organization transformation? Most people would assume it was the senior group. However, that group had become so frustrated with the intransigence, inflexibility, and mind-numbing routine of the agency that they welcomed and even sought change, almost anything, in the hopes that the working environment would improve. Even more surprising, perhaps, is that the "newbies," having just gone through a traditional indoctrination training period, were almost brainwashed to the extent that they resisted any hint of derailing what they had just learned before they could start applying it in their new jobs.

Many, if not most, organizations cannot function at peak expectations without a high degree of trust, not only among the individuals within the organization but also between themselves and other organizations with which they interface strongly. The importance of the trust factor borders on the ridiculous from a rational point of view, but from an emotional point of view it is completely understandable. Why shouldn't competent individuals who are not known to you be trusted to provide useful information to help the organization perform its mission? Because doing so might create vulnerabilities that put you or your work in jeopardy. Few individuals are willing to jeopardize their careers by trusting outsiders, lest something bad happen; for example, closely held information about intent and operational procedures might come to light that would lead to criticism of the organization from either competitors for the same work and/or budget dollars or, heaven forbid, objective sources who might offer better ways of doing things.

Here's a typical situation that can lead to considerable frustration. Suppose you are trying to introduce a new person into the organizational mix by allowing him or her to start attending meetings as a fly on the wall to listen, primarily, to gain an appreciation of the group and what it is trying to accomplish. The thought is that this person, once assimilated and brought up to speed, will be able to become a more active participant and add value to the group's endeavors because of that individual's particular knowledge and skill. However, this typically is not allowed to happen unless some respected and authoritative member of the organization runs interference by selling the idea to the participants. Otherwise, the outsider will be marginalized—either not allowed to attend to meetings or allowed to attend but be destined to fail because no one will say anything significant while that person is present. Groups accept new members more readily if a trusted and respected member vouches for the newcomer's trustworthiness. However, at least two other factors facilitate acceptance into the group: (1) if the newcomer starts injecting good ideas based on his or her knowledge and the group begins to recognize the value of the ideas and/or (2) if, over time, the group begins to get comfortable with the new person.

Most leaders of organizations with a hierarchical management structure likely have attained their positions through a series of assignments with increasing responsibility. On their way up the career ladder, the more successful will have exhibited a can-do but risk-averse mindset, paying lots of attention to their superiors' intentions. This is especially true in command-and-control organizations, particularly those of the military or those modeling their structure after the military. As a class, these leaders are not used to experiencing failure in their professional or personal endeavors. As their level of responsibility increases and they become immersed in more complex environments, situations become more difficult, and they may find themselves less effective in achieving their goals through their dominant, top-down, directive, or sometimes heroic leadership styles. In a way, the Peter Principle takes over. Such leaders usually will not accept much of the share of responsibility for failure, but they will become quite defensive if questioned and will blame the problem on others using some convenient rationale (Argyris 1991). This hinders their own learning, and that of the organization, by reinforcing a culture of blame and sabotaging the frank surfacing and exchange of critical negative information that is necessary for building interpersonal and organizational trust. Argyris (1991) talks about the need for "double-loop" learning, a desirable mindset that encourages honest and objective self-assessment in the spirit of continual improvement where informed mistakes are not only tolerated but even rewarded. Heath and Heath's (2007) "curse of knowledge" is another factor that hampers senior decision maker learning. These leaders know so much about how they got to where they are, they are less likely to consider alternative methods when confronted with what might seem to them to be insurmountable (extremely complex) problems. Among other things, this is not conducive to organizational learning (Argyris 1994; Vince 2002).

Enlightened decision makers encourage their staff to challenge themselves and management, instead of trying to create a happy environment that leads to employee dependency on management (refer to Covey's codependency anathema) and a disengagement from their own proactive examination, assessment, and reflection on the effectiveness of the organization (Argyris 1994). Further, such decision makers encourage their colleagues to challenge them. "Good strategists make sure that their conclusions can withstand all kinds of critical questioning" (Argyris 1991).

We believe in a hybrid organization, in which hierarchy has a role in helping to create the organizational vision and ensure that people are empowered to carry out the organization's mission. Indeed, societies, let alone organizations, have been known to collapse without a sufficiently strong hierarchical sub-structure (Tainter 1988). However, the individual and personal networks that form and re-form dynamically within organizations are where the real communication occurs and the work of the business is carried out (McCarter and White 2007).

Overarching Fragility Concern

Every organization must operate in the larger context of civilization in general and within a particular society. Complex situations can be fragile and vulnerable to potentially disruptive, if not catastrophic, effects (e.g., from a pandemic flu) (MacKenzie 2008). On the other hand, healthy complex systems are generally more robust. Thus, one should try to build more redundancy and diversity into the system to help enable more robust capabilities. The more complex the environment, the more we must depend on others for necessities—those things that if supplied allow us to specialize and to create value-added contributions to our organization, society, and civilization, in turn. It is human nature to discount the possibility of devastating, traumatic events such as nuclear war, pandemic flu, and financial meltdowns like what the world witnessed in late 2008 (Jameson 2008; http://en.wikipedia.org/wiki/Financial_crisis_of_2007%E2%80%932010) and again in August 2011. We rationalize that these kinds of events are too unlikely to worry about and even if they do occur, they will not affect us—only others. However, if we think more rationally and realize that these "black swan" events (Taleb 2007) can occur, and with a probability much more likely than something on the tail of a bell curve (i.e., with probabilities that follow the power law), there is cause for grave concern. A pandemic, for example, as MacKenzie (2008b) so aptly describes, can bring everything to a screeching halt. So what do we do? We can try to utilize or introduce certain patterns or features that increase the robustness of complex systems or enable a complex system to degrade or fail gracefully. Other than that, we can only try to be more realistic and consider the possibility of black swan events when we are planning. It is almost more important to think about failure modes, and what we would do when things go wrong, than to concentrate on finer and finer grained details of processes that assume everything is going according to plan.

Recognize That Complex Systems Can Do Better than We Can

We must bring humility to the study of complex systems. Buchanan (2008, 28–31) argues that non-human elements of complex systems can self-organize to create more effective and efficient solutions than people can:

> The wider lesson is that we just can't trust our intuition when it comes to the super-complex systems that we depend on today. We may never learn exactly how to control these systems in the traditional fashion and the best way to cope may be by learning new principles for letting them manage themselves. Engineering isn't just about solving problems any more, but building systems that can solve their own problems. Being in control, it seems, may increasingly demand being a little out of control. On the other hand, attempting to realize this can lead to other problems such as the people's

wariness of using any idea of artificial intelligence (AI) to help manage the work of robots (bots)—let alone their work! What might be even worse in this regard is to contemplate allowing bots to create other bots. So a certain amount of care and education is needed to further this concept of self-help and self-organization.

Chapter 2

The Nature of Being Human

B. G. McCarter

One of the most complex dynamics we face is that which involves ourselves. Complexity arises from many dynamic interactions among different variables. People interacting with each other continually on a daily basis, especially globally, results in unpredictable and uncontrollable scenarios. Historically, leadership has been seen as the ability to influence or control groups of people for certain desired ends. In organizations and cultures, particularly in the foreign policy arena, which can impact the likelihood of armed conflict, this elusive quest for what makes the best leader has resulted in a tremendous amount of research and investments of money and time.

The advent of the scientific method and the general beliefs of the Age of Enlightenment furthered man's attempts to understand and influence those around him. Humans looked for linear cause and effect in governing people's behavior. They looked for rational and logical reasons for the way people behaved. The behaviorist movement in psychology spawned a variety of management techniques that continued to look at humans as functioning as predictable machines. Inherent complexity, however, means that the interaction of variables—in this case, humans—results in emergence: unpredictable, uncontrollable results. Man is definitely neither a logical nor a rational machine. We are human, and what that means has been debated by philosophers for centuries.

Neuroscientists have discovered a wealth of information about the brain during the past several decades in their attempt to try to understand how the mind, and subsequently our behavior, emerges from the interactions of the various parts of the brain, our bodies, and our environment. Our thoughts are influenced by a plethora of variables ranging from our genetic encoding, to our individual neurochemistry, to cultural influences. These thoughts influence our behavior, and our behavior affects our interactions with others. Ultimately, these interactions can lead to unforeseen events. The notion of leadership becomes much more complicated, indeed.

Mind and the Brain: An Overview

Antonio Damasio, in his book *Descartes' Error* (Damasio 1994), points out that our ability to reason has evolved as an extension of our basic emotional system, which existed to help us survive in our environment. As a result, emotion plays varying roles within our decision-making processes.* As images flood the mind as we try to make decisions, emotion facilitates our holding these images until we make a decision. As demonstrated by some neurological conditions, when emotions

---

* Another thoughtful departure from Descartes credits emotion as key in the development of abstract thinking and the meaningful embracing of symbols (Greenspan and Shanker 2004). These authors also argue how autistic children develop rational thought much more slowly, if at all, because of the lack of emotional connection with others. They also claim Chomsky and others were wrong in how the use of language suddenly appears in human behavior without a direct relationship with emotional development.

are left out of the decision-making process, our reasoning is impaired. Emotion plays a role in our intuition: that magical process by which we are able to make rapid decisions seemingly without any logical analysis (Barch and Morsella 2008). The subconscious or "non-conscious" gathers data points over time and stores them in the mind, making lateral connections as we experience life and establish patterns.* Emotion helps to facilitate that "ah-ha" moment when the data points are not consciously connected, and we just "know" what we have to do. It is the same mechanism that helped humans in earlier times survive in dangerous environments.

Damasio also points out through his research with neurology patients that the brain systems, while "jointly engaged in emotion in decision-making are [also] generally involved in the management of social cognition and behavior" (Damasio 1994, xiii). This finding points to a connection between individual neurobiology and social and cultural factors; for example, the role of chemicals such as dopamine and oxytocin, which can change our feelings, thus our thoughts, and hence our behavior (Baskerville and Douglas 2010).

Patients who, as adults, suffered injury to the parts of the brain controlling emotion demonstrated that emotions were required to execute proper social behavior, whereas those suffering injury to the same brain areas early in life "showed that emotions were also needed for mastering the know-how behind proper social behavior" (Damasio 1994, xiv). These

---

* The huge role of the subconscious, which is responsible for the vast majority of our brain's workings, is also emphasized by Brooks (2011); he also advocates that our emotions assign values to alternatives, which goes far in how we ultimately make decisions, albeit using some rational thought. Another point to ponder: "Rather than seeing the world as it is, you see it through a veil of prejudice and self-serving hypocrisies. ... The problem is that our biases—which form and solidify in childhood and early adulthood—operate below the radar, in our subconscious. ... confronting people with new information that contradicts their beliefs more often than not ends up hardening their position" (Lawton 2011, 37–38). Thus, it is important to be aware of subconscious prejudices in ourselves and others to understand our mutual fallibility, to bring some humility to the discussion, and to help reach understandings of mutual benefit.

studies point to a connection between neurobiology and culture. If the emotional part of the brain is damaged, depending on when the injury occurs, the patient may either know the rules of social behavior but not be able to enact them or not even know there are rules for social behavior.

The absence of emotion also compromises the ability to make rational decisions. For patients who experience electroshock therapy, a common symptom is the inability to make any decisions at all. Patients are able to analyze all the factors that would go into making a particular decision, yet they are unable to actually make a decision. This is also seen in patients who have suffered damage to the emotional part of the brain through other means.

Emotion helps people make decisions in an uncertain environment, wrestle with moral judgments, and make plans in relationships as well as for the future. There is not one single area of the brain that is used in making decisions, but rather a complex organization across several brain systems working from a high level to low level, involving the prefrontal cortices to the hypothalamus and brain stem, working together to enable us to reason. "Emotion, feeling, and biological regulation all play a role in human reason" (Damasio 1994, xvii).

In addition, the view of reality people construct in their mind for themselves is subjective and based on the neural signals sent from the body to the brain. How we emotionally experience life is determined by chemical and neural processes influenced by our bodies. Do we feel happy, sad, angry, or hungry? Do we trust others? These emotions are influenced in part by neurological and physiological effects. Evolution determined that the mind had to be mostly concerned about what was happening to the body; otherwise, humans would never have survived. The body gives the mind constant up-to-date feedback about what is happening to it. With this feedback as a given the mind is then able to do other things as well.

[This] idea is anchored in the following statements:
(1) The human brain and the rest of the body constitute an indissociable organism, integrated by means
of mutually interactive biochemical and neural regulatory circuits (including endocrine, immune, and
autonomic neural components); (2) The organism
interacts with the environment as an ensemble: the
interaction is neither of the body alone nor of the
brain alone; (3) The physiological operations that we
call mind are derived from the structural and functional ensemble rather than from the brain alone:
mental phenomena can be fully understood only in
the context of organisms interacting in an environment. That the environment is, in part, a product of
the organism's activity itself, merely underscores the
complexity of interactions we must take into account.
(Damasio 1994, xx–xxi)

The mind develops from the interaction of the body, the brain,
and the environment.

Basics of the Brain

One of the most well-known cases of a person who suffered
injury to the brain and survived was a man named Phineas
Gage. In 1848 Phineas Gage had a railroad-building accident
that resulted in a steel spike going through the frontal lobes of
his brain. That accident changed his entire personality, including his likes and dislikes, his ability to plan for the future, his
ability to make good choices, and his social interactions. The
change in Phineas Gage pointed to something in the brain
being responsible for human behaviors that were often considered to be under the realm of free will. Prior to the accident,
Gage had been well thought of, responsible, and well man-
nered. After the accident, he was ill tempered, foulmouthed,

and irresponsible, but people could not understand that he was not actively choosing to behave so horribly. Gage's case pointed out that being able to get along in society, make ethical judgments, and make decisions to secure one's survival requires not only understanding cultural rules and having the ability to plan, but also that specific brain systems are functioning properly. Human conscience as well as culture seem to depend on a particular area of the brain.

"[The] brain is a super system of systems, i.e., complex system [in the sense of Chapter 1]. Each [sub-]system is composed of an elaborate interconnection of small but macroscopic cortical regions and subcortical nuclei, which are made of microscopic local circuits, which are made of neurons, all of which are connected by synapses" (Damasio 1994, 30). Subsequent neurological research into Phineas Gage's situation led to an understanding that the ventromedial region of the frontal lobes is critical for normal decision making. Other neurological cases of damage to the prefrontal part of the brain also have resulted in radical personality changes. These included the inability to reason, learn from mistakes, plan for the future, or make decisions. It seems that in the cases studied, the brain structures destroyed were those necessary to have reasoning result in decision making. There is a difference in how society views people who have suffered diseases of the brain versus diseases of the mind. If you suffer a stroke that causes paralysis of your limbs, you are looked at as a tragic victim. But if part of your brain is damaged that affects your conduct in society, then you are assumed to have a lack of willpower or a character flaw. In reality, the brain and the mind are inseparable. The mind emerges from a complex interaction of multiple parts of the brain. Our emotions, understanding of social conventions, and ability to make decisions rationally are all affected by this interaction. Patients with frontal lobe damage also seem to have difficulty generating estimates based on incomplete knowledge. The mind avoids making errors by using probability and top-down analysis to estimate. Damage to the frontal lobes also damages one's emotions. Individuals who suffer this type of damage to the brain are aware of the feelings they should be having based on circumstances experienced, but they have no true emotional response. They know what they are supposed to feel, but they do not have the feelings.

Research shows an interaction of brain systems that reveal the underlying processes of emotion, feeling, reason, and decision making. Emotion underlies the ability to reason. Emotions can cause problems in making decisions, such as when one is angry, but a lack of emotion leaves one unable to make good decisions or even make any decisions at all. Somewhat surprisingly, however, one is still able to do some reasoning without emotion. Damasio points out that studies of patients who suffered damage to the frontal lobes during childhood or adolescence showed social behavior problems, but their intelligence was not affected. Access to emotions was also demonstrated to be lacking as well as a initiative. Rewards and punishments did not influence behavior, and patients were neither happy nor sad. "They are bereft of a theory of their own mind and the mind of those with whom they interact" (Damasio 1994, 58). These patients were not able to generate patterns of behavior based on cause and effect nor were they able to predict and generalize from their own actions how others might feel and behave. They had no idea of their own social role. Their emotions seemed flat and the drive to respond or act was stifled. They were less creative and less decisive.

Other demonstrations of effects on the frontal lobes of the brain include leucotomy, electric shock therapy, psychotropic drugs such as thorazine, and anosognosia, where having suffered some sort of damage to the brain, the patient will deny that anything is physically wrong despite obvious demonstrations of having limbs that do not function properly.

In anosognosia, the body's sensory system malfunctions and there is a lack of emotion. Because these patients cannot establish a theory of mind, they are unable to foresee consequences. Just like prefrontal patients, anosognosic patients have difficulty making proper decisions regarding personal and social issues. It seems they have malfunctions in reasoning and decision making as well as emotion and feeling.

Research points to areas of the brain that impact both reasoning and decision-making processes: (1) ventromedial prefrontal cortices and the amygdala, where emotion and reason intersect; (2) the somatosensory cortices in the right hemisphere, involved in the processes of basic body signaling; and (3) other regions in the prefrontal cortices beyond the ventromedial area. "In short, there appears to be a collection of systems in the human brain consistently dedicated to the goal-oriented thinking process we call reasoning, and to the response selection we call decision making, with a special emphasis on the personal and social domain. The same collection of systems is involved in emotion and feeling, and is partly dedicated to processing body signals" [Damasio 1994, 70, 71 (refer to Figure 4.4)].

According to Damasio, a large number of serotonin receptors reside in the ventromedial sector of the prefrontal cortex as well as in the amygdala. Serotonin is one of the main neurotransmitters that affect social behavior. In primates it inhibits aggressive behavior and favors social behavior. This research points to an interactive connection between the ventromedial prefrontal cortices and the amygdala, relating these regions to social behavior that is affected by flawed decision making.

Again, the brain is a complex system where multiple areas interact and with various neurochemical influences. If one aspect of the brain system is affected, this can modify the operation of other systems in the brain. Like most complex systems, if you change one variable you can have dramatic impact on the whole system.

The systems Damasio discusses are involved in the brain's reasoning processes, especially planning and deciding. One subset of the systems involves personal and social behaviors. This relates to what we usually consider rationality. These systems also are involved in the processing of emotions, as well as the ability to hold in the mind images not present but necessary in order to make decisions. Thoughts and knowledge come from the integration of multiple areas in the complex system of the brain. There are many processes active in our brain about which we are never aware. Our subconscious or non-conscious is always taking in information. We only become aware of that information when our non-consciousness deliberately draws attention to it by alerting our consciousness. And like all living systems, our brains and minds are constantly changing and in flux.

Mind, Body, and Environment Interaction

As stated before, the mind is an emergent phenomenon resulting from a complex system of interactions. There is also a complex interconnected interaction of biochemical and neural circuits. One interactive route is through the sensory and motor peripheral nerves, which carry signals back and forth between the body and the brain. Signals transmit into the brain through the spinal cord or the brain stem, and signals relay from the brain to the body through the autonomic and musculoskeletal nervous systems. The autonomic nervous system derives from the older areas of the brain: the amygdala, the cingulate, the hypothalamus, and the brain stem. Another route is through the bloodstream through such chemical signals as hormones, neurotransmitters, and modulators. These mechanisms all influence the brain's operations. The brain is influenced by the body through the manufacturing and releasing of chemical substances into the bloodstream. In a complex feedback loop the brain and body interact with the environment. According to Damasio, "having a mind means that an organism forms neural representations which can become images, be manipulated in a process called thought, and eventually influence behavior by helping predict the future, plan accordingly, and choose the next action. ... [This is the] center of neurobiology ... the process whereby neural representations, which consist of biological modifications created by learning in a neuron circuit, become images in our minds; ... [which affects] invisible microstructural changes in neuron circuits ... [which] becomes an image we each experience as belonging to us" (Damasio 1994, 90). The purpose of this complex process of the brain is to help the body survive in its environment.

Instead of having direct feeds of sensory signals to the brain, the neural systems host a complex series of multiple, parallel, and converging streams. The streams never terminate but are constantly projecting forward, producing a feedback loop into the brain system that allows signals to go forward and backward. The neuron groups are complexly interconnected with no direct input roots. Damasio feels that they are complex in order to allow us to momentarily construct and manipulate images in our minds. Thus we recognize the incoming signals as concepts and assign categories for them. Categorizing the concepts, in turn, allows us to establish strategies for decisions.* We can then choose a physical response to enact. This system of systems allows us to make consciously willed decisions.

The Mind and Emergence

There is no one area of the brain able to process simultaneously all the sensory information the body is collecting from the environment, such as through touch, sound, sight, as well as temporal and spatial relations. Various sensory experiences come to the brain in a variety of interactive locations, not just one. The mind emerges from the complex interaction of multiple parts

---

* Harold Klein and William Newman have used a strategic decision-making tool, called SPIRE, in a complex environment. The ideas on which they rest SPIRE reflect the neuro-anatomy relationship to decision making in this chapter. There is a great deal of overlap, and Klein and Newman have built their decision-making tool on that predication.

of the brain. According to Damasio, "damage to those higher-order convergence regions, even when it occurs in both hemispheres, does not preclude 'mind' integration at all, although it causes other detectable neuro-physiological consequences such as learning impairments" (Damasio 1994, 95). Synchronization of the various parts of the brain requires approximate timing for aspects of the mind to emerge. The mind is integrated through time. States of confusion caused by blunt trauma to the head or through mental illness are often a result of a malfunction of the time mechanism in the brain. To bind time, an organism needs the powerful mechanisms of attention and working memory. Each sensory system in the brain seems to be equipped with both of these mechanisms. Prefrontal cortices and some limbic system structures are essential to the processes of global attention and working memory.

Images and Knowledge: What Is Reality?

In *Descartes' Error*, Damasio reveals that "factual knowledge required for reasoning and decision making comes to mind in the form of images" (Damasio 1994, 96). The images the mind generates come from two sources: perceptual images and recalled images. Perceptual images come from a variety of sensory systems, including landscapes that we see with our eyes, music to which we listen, textures that our touch experiences, or the abstract symbols of words we read. Recalled images are memories of past experiences. These also include images of plans we have made for the future. All of these images are constructions of the brain. And because every living creature's makeup is different, every living creature's perception of reality is different.* Research shows that the images

---

* Although this notion is readily accepted by most people, the idea that people can *choose* their own reality, a fascinating concept (Zeland 2008, 10, 11), is more controversial.

we perceive of the environment within and around us arise from the interconnections of a variety of areas of the brain. In order for the images to register in the mind, those neural representations "must be correlated with those which, moment by moment, constitute the neural basis for the self" (Damasio 1994, 99). One has to have a sense of self, which Damasio feels is "a perpetually re-created neurobiological state," for the images to register as *our* images (Damasio 1994, 99).

It is a poor analogy to say that images in our memories are stored like files in a filing cabinet. Because many different parts of the brain interact to create an impression of an image, we do not get an exact representation but rather an echo of the original event. Similar to an orchestra playing a symphony, any sensory responses to an image or memory result in different firings of neural connections in different parts of the brain to create the whole experience. Recall of an image means these different neural connections need to be activated or "strummed" again as they were the first time. A complex sequence of neural firing patterns re-creates an interpretation of the original image. It is an echo of the original image and subject to discrepancies in the exact pattern being activated. Those discrepancies can cause the memory that is recalled to be less than exact. Although there can be inaccuracies in memory, using so many different areas of the brain to create patterns that reconstruct an interpretation of the original image leads to the brain's greater capacity in accessing and storing imagery.

In creating a theory of mind to understand how others may be behaving or what they may do, one does not necessarily have to have had direct experience in a similar situation. Mirror neurons in the brain (http://en.wikipedia.org/wiki/Mirror_neurons) allow us to understand the intention and movements of others even though we have not experienced something physically ourselves. We are able to grasp some things in a more intuitive fashion, understanding an experience as an image rather than as a verbalized concept. Human thought is made up largely of images. Damasio points out that most of the words we hear inside our own heads exist as either auditory or visual images in our consciousness. He asserts that if they did not become images we would not be able to understand or know them. Knowledge is stored as images, and our subconscious or non-conscious self is constantly processing these images even when our conscious self is not even remotely aware of that process. Damasio refers to experiments that demonstrate images processed by our non-conscious self can nonetheless affect our thought process and even later emerge into our consciousness.

Why does the mind operate this way, using imagery as the main content of our thoughts regardless of the sensory process that created it? The flexible and adaptive interconnections of different parts of the brain enable an organism to record experiences and responses to them, evaluating those responses in shaping basic preferred response patterns to help ensure that organism's survival. Modulator neurons help to exert influence on an entire set of circuits in the brain that regulate body function in order to change an organism's behavior to maximize survivability. Modulator neurons control an organism's internal biochemical systems by distributing such neurotransmitters as dopamine, norepinephrine, serotonin, and acetylcholine. The brain continues to develop from infancy to adulthood, influenced by the evolving body and its interaction with the environment. The mind is adaptable and flexible in response to an unpredictable environment because it has the ability to learn. Brain circuits continue to change over time, reflecting the changes an organism experiences. This enables us to continue to recognize ourselves in the mirror as we age over the course of years.

Innate Dispositions for Survival

The autonomic response of fight or flight results from the neurocircuitry that controls hard drives and instincts to help ensure survival in the environment. Damasio asserts that behaviors derived from drives or instincts contribute to our survival by influencing decisions that reduce the probability of harm to us. Emotions and feelings are essential components of those powerful drives and instincts. Although we may not be aware of basic regulatory mechanisms such as circulating hormones, the number of blood cells in the body, or potassium ions, we are aware of our instincts, such as feeling hungry and eating.

Current instincts can be modified or individualized based on experiential feedback. There is a complex interaction between the brain, mind, and body that is inseparable. Complex feedback loops regarding chemical regulation such as hormone production demonstrate this interconnectedness. Another example of the brain, mind, and body interaction is when chronic mental stress results in an overproduction of the chemical calcitonin gene-related peptide (CGRP), which coats immune-related cells and makes the body more vulnerable to infection. Bereavement can result in depression of the immune system as well. Yet another (positive) example is oxytocin, a naturally produced chemical substance that increases feelings of love and trust. It affects both the body and brain and is manufactured in both the brain and body. Not only does it facilitate trust in social interactions and bonding between sexual partners, it also acts on the body by relaxing muscles during childbirth. Finally, primitive brain components interact with other body parts and processes to enhance survival capabilities. Damasio sums up this complex interconnectedness in the following passage:

> The hypothalamus, the brain stem, and the limbic system intervene in body regulation *and* in all neural processes on which mind phenomena are based, for example, perception, ... learning, recall, emotion and feeling, and ... reasoning and creativity. Body regulation, survival, and mind are intimately interwoven. (Damasio 1994, 123)

Beyond the Non-Conscious

In order to survive in a complex and ever-changing environment, humans must rely not only on evolved non-conscious instinctual responses, but also on cognitive survival strategies developed and taught in societal environments. These cognitive strategies require a deliberate reasoning process. Reason is able to provide a check for emotion, and thus healthy individuals do not act out in a dysfunctional manner when feeling extreme levels of such emotions as anger, sadness, or hunger. Normal healthy individuals do not kill people when they become very angry, nor themselves when profoundly sad, and they do not necessarily go on a feeding frenzy when they are extremely hungry. The complex feedback loop between the mind, behavior, and the environment helps to allow reason to act as a governor on emotions. Freud called it the superego whereas philosophers today refer to memes. Much like the mind and its emergence from the interaction of the various parts of the brain, so too does culture emerge from the interaction of the behaviors of individuals. Just as we cannot deconstruct the phenomena of the mind to various individual parts of the brain, we cannot reduce culture to a basic genetic pattern found in an individual. To better understand cultural dynamics, Damasio asserts that we need to combine research in the social sciences with general biology and neurobiology.

Individual behavior is affected by an interaction of processes within the mind and body and experiences in the environment. Group behavior emerges from the interaction of individuals, whose behaviors are also affected by the previous three factors. Social conventions and ethical rules can help shape instinctual behavior, which, in turn, can help individuals be adaptable and flexible in ever-changing complex environments. Most societal and cultural rules were developed to facilitate the survival of individuals as a social group. Damasio brings up the question of whether this means altruism and free will are just instincts and do not really exist.* Although people tend to follow the behavioral customs and rules of their culture in order to maximize survivability, this does not mean that they always agree with or follow those conventions. The interaction of neurobiology and society affects how we feel about things, but individuals can act contrary to culture in regard to moral decisions.†

Uncertainty in complex situations brings into action systems located in the neocortex of the brain, a more modern, evolved section of the brain, as opposed to the amygdala, which is part of an older system. According to Damasio, "there is evidence for a relation between the expansion and sub-specialization of the neo-cortex, and the complexity and unpredictability of environments with which such expansion permits individuals to cope" (Damasio 1994, 127). Rational decisions on how to react in an ever-changing complex environment require an integrated interaction between the neocortex, the evolved modern brain, and the hypothalamus, the older seat of emotions, drives, and instincts. As demonstrated with patient cases such as Phineas Gage, rational thought in the neocortex does not exist without interaction from the subcortical hypothalamus and amygdala, the seat of emotions. Our system of rational thought is not just an additional system in the brain built on top of an older existing system that regulates our biological functions, but rather, it evolved from and in concert with that older system. There is an interactive dialogue at play between the two systems that allows rationality. Emotions and feelings are the bridge between rational and irrational thought.

---

* This is a fertile topic; for further discussion, refer to Ozinga (1999) and Zaki (2009).
† Trevino (1986) offers a rather good treatment of this subject in the context of moral decision making in organizations, a thrust that is quite relevant to the present book.

Emotions

Emotions are tied to expressions of physical symptoms. They would be hard to describe or understand otherwise. For instance, the emotion of rage is accompanied by rushing adrenaline, tightened muscles, heavy breathing, an animated face, flushing of the face, dilation of the nostrils, and clenching of teeth. Emotions are tied to biological processes. They also are tied to mental processes that in a feedback loop can trigger physiological and biological processes. Emotions play a role in communicating meaning to others. Rational thought modifies our preset emotional responses through evaluation and deliberation. In effect, rational thought can affect the basic machinery of our emotions.

Some emotional reactions to sensory perceptions are hardwired into us at birth in order to facilitate survival. For instance, the size of an object might relate to certain animals, span might indicate winged predators, motion might reveal a reptile nearby, and sounds might relate to hidden growling in underbrush. Our emotional response prompts us to react quickly. But if it is a knee-jerk survival instinct, why do we need to be conscious of our emotions? Damasio (1994) surmises that by knowing the specific threat prompted by the emotion, we have greater control over the types of behavior with which we react. This allows us to generalize our knowledge and established pattern recognition through learning based on interactions with our environment. These primary emotions jump start the process.

Pattern recognition from experiences with our environment results in emotions that trigger physiological responses. Certain situations become paired with certain emotional responses. These emotions become acquired rather than innate and are referred to as secondary emotions. The pattern of responses individuals develop are unique to the individual. This is another example of how individual views of reality are constructed. Secondary emotional responses need the primary emotions in order to express themselves.

Damasio traces this complex interactive process among the mind, body, and emotion when he explains that "emotion is the combination of a *mental evaluative process,* simple or complex, with *dispositional responses to that process,* mostly *toward the body proper,* resulting in an emotional body state, but also *toward the brain itself* (neurotransmitter nuclei in brain stem), resulting in additional mental changes" (Damasio 1994, 139). It is interesting to note the studies Damasio examined that reveal the separate nature of motor control and emotion. It seems motor control related to an emotional response is located in a different part of the brain than the voluntary expression of that same muscle pattern. For instance, this is the reason some people have difficulty smiling naturally for photographs. Smiling for the photograph is a willful control as opposed to a natural emotional effect.

Current understanding of the complex interactions of the various parts of the brain with the body and the environment have made passé previous diagrams of the human brain as a rigidly compartmentalized structure. The brain functions much like a highly engaged network system as it processes in real time what is happening in the body. During an emotion the body undergoes pervasive changes. Feedback loops involving an organism's emotional state are present in both the neural and chemical networks.

Interaction with our environment can trigger emotions, which can then affect the body, which in turn can produce chemicals and hormones, which can affect our emotions. The chemicals produced can affect how the neural signals are processed. Damasio (1994) surmises that this may be why, historically, chemical substances have played an important part in many cultures and why our society is wrestling with its current drug problems. Emotions can affect not only the physical body (such as when depression suppresses the autoimmune system), but also rational thought. For instance, depression can banish rationality, leading to suicidal thoughts. Feelings are based on a combination of neural imagery, feedback loops of bodily responses to an experience, the perception of the body state, and thought processes associated with that experience.

Research has demonstrated that because of the interactive nature of these feedback loops, body states can cause feelings. It's all about attitude. You can make yourself happy just by believing so. Experiments have shown that even incompletely composed happy facial expressions used in tests resulted in the subjects experiencing happiness. Conversely, angry facial expressions resulted in the subjects feeling angry even without experiencing initiating events. This may well be related to the mirror neurons in the brain, which help people understand the actions of others and develop a theory of mind. The research demonstrated that fragments of body patterns representing a particular emotion were enough to produce the feeling. The emotion was triggered, yet at the same time, the subjects were conscious that they were neither happy nor angry at any particular thing.

Like all complex systems, brain and body states are not predictable, and the results of their interactions are not reproducible. Emotion is induced by not one route, but two: the neural route and the chemical route:

> [T]he brain is not likely to predict how all the commands—neural and chemical, but especially the latter—will play out in the body, because the play-out in the resulting states depends on local biochemical contexts and on numerous variables within the body itself which are not fully represented neurally. What is played out in the body is constructed anew, moment by moment, and is not an exact replica of anything that happened before. ... [T]he body states are not algorithmically predictable by the brain, but rather that the brain waits for the body to report what actually has transpired. (Damasio 1994, 158)

Reason and Decision Making

With all the various processes interacting in such a complex fashion, how does one feel an emotional state? The process is not completely understood, but aspects are being revealed. To pair emotions with a person or an event, the brain has to have some means to represent a connection between the person or event and the resulting body state. When we make incorrect links, it often leads to superstition or phobic behavior. Again, emotions, feelings, and body states are all interconnected through complex feedback loops. The constant process of creation and connection building is what reason and decision making are all about. The present does not exist if one is too busy examining the past in order to plan for the future. To be able to reason or decide usually implies a person has knowledge: (1) of the situation; (2) about possible options for responses; and (3) of potential immediate and future consequences of those responses.

Knowledge in memory can be made accessible to consciousness in both verbal and nonverbal versions simultaneously. One example of this is through a biochemical means: a person's blood sugar level drops, triggering neurons in the hypothalamus, which in turn induces a hunger state to encourage the person to eat, for instance. Another example is the stimulus response patterns, such as when one avoids a falling object. Pattern recognition from experiences in the past provides a ready strategy to act upon for survival. A third example is conscious deliberate choice; for instance, choosing a job, deciding whom to vote for, or deciding whether to forgive somebody. A different kind of conscious choice involves the more practical/complex area of solving a puzzle, painting a picture, or building a bridge.

Because of the prevalence of complexity and uncertainty in today's world, it is not easy to make reliable predictions to guide what actions one chooses. To establish a management strategy, however, a host of possible actions and outcomes must be generated. To complicate things further, to make that final decision one most hold all these ideas in mind, testing them against the goals of the organization as well as listening to one's intuition that has been honed from years of experience.

What is the process by which we reason? Our minds are flooded with images relevant to the situations we face, helping us compose possible response options and outcome scenarios. Questions are brought to mind by the images generated. The mind is not blank when we start to reason. One commonly accepted perspective of how we reason is cool formal logic, as depicted by the *Star Trek* character Mr. Spock. One begins with a cost/benefit analysis, supposedly keeping emotions out of the process, inferring what is good and bad while generating a list of possible outcomes and evaluating each. It is a subjective process that is also lengthy. It is difficult to hold in one's mind all the multiple contingencies and whether something is good or bad, although pencil and paper does help the process. Error can be introduced by a lack of personal knowledge of all possible outcomes and consequences.* Damasio (1994) wryly notes that the cool strategy used by the philosopher Kant resembled more the way patients with prefrontal lobe damage made decisions as opposed to the way healthy functioning individuals make decisions.

Damasio (1994) puts forth a second perspective that he calls the somatic marker hypothesis. This involves a physical reaction to pattern recognition. Unpleasant physiological responses to negative outcomes are paired with certain actions. Often there is a fleeting emotional response that triggers the body's alarm system to possible negative decisions. This may cause a person to reject instantly a possible negative action and is thought to be a survival mechanism. It

---

* Similarly, traditional economics and its theoretical underpinnings are based on the assumption that people have perfect knowledge in all things relevant to making economic decisions. However, this does not square with the complexities of economic behaviors in the real world (Beinhocker 2006).

utilizes intuition in that the emotions and feelings are paired through learning experiences to help the person predict future outcomes. This type of decision-making process often comes into play in social situations that require individuals to form adequate theories of mind for not only themselves but also for others with whom they interact. We predict what others are going to do based on understanding how we would behave. These predictions are affected by the somatic markers. It is important to note that the physical response to situations and the resulting lessons learned from experiences are not necessarily the same for different people. This contributes to conflict in interactions due to our inability to understand each other's individual views of realities.

What then determines "truth"? Individuals construct their own truth within their view of reality. Subsequently, each view of reality—our beliefs, feelings, and intentions—is influenced by a host of variables seen and unseen in the complex environment in which we live. Some of these factors result from the complex neurophysiological interaction of the mind, body, and environment with each other; other factors relate to experiences growing up in a family; and others are influenced by societal and cultural rules and regulations. Individual biology as well as the culture affect reasoning processes. Despite that, however, individual free will exists, and demonstrations of individuals performing actions contrary to their biology and culture are seen periodically.

The origins of somatic markers seem to be genetic. Neural responses are paired with primary emotions to help navigate the complex signaling that regulates personal and social behavior. They also seem to be part of the education and socialization process. Neural processes are used to aid rational decision making and, according to Damasio (1994), are based on our secondary emotions. For these somatic markers to function properly, both the brain and the culture in which the individual is immersed must be normal. When either is too dysfunctional severe consequences can be seen. When the brain is defective, a host of varying levels of mental illness may be present. Extreme examples include sociopaths and psychopaths, who are extremely rational and calculating yet also unfeeling and uncaring. The psychopathic state results from dysfunctional brain development or abnormal chemical signaling in early brain development. Extending this concept, a defective culture can negatively impact otherwise healthy human adults. Damasio (1994) offers several examples: Germany and the Soviet Union in the 1930s and 1940s, China and its Cultural Revolution, and Cambodia with its Pol Pot regime.

Somatic markers are "under the control of an internal preference system and under the influence of an external set of circumstances which include not only entities and events with which the organism must interact, but also social conventions and ethical rules" (Damasio 1994, 179) These markers are learning based and are constantly changing as a result of interactions with the environment. We acquire somatic marker signaling through the prefrontal cortices. The prefrontal cortices receive signals from all of our sensory systems, which help us make sense of the world in which we live. These signals create images in our mind to re-create an impression of the original experience. Included in these sensory systems through which the prefrontal cortices receive signals are the somatosensory cortices, which continuously represent past and current body states. These are derived from the perceptions we have of our environment, our thoughts about that external environment or physical events encountered, and our bodies. The prefrontal cortices are aware of almost all activities in our mind or body at any moment.

The prefrontal cortices receive signals related to an organism's survival—our innate preference—and thus, the prefrontal cortices are part of the reasoning and decision-making process. They are involved in the categorization and classification of our life experiences based on a wide variety of characteristics. The prefrontal cortices categorize our experiences in real

life as well as the contingencies associated with those experiences. They establish our reactionary dispositions—patterns of
behavior standing at the ready in case we need them again in
the future. As a result, they help shape our view of reality. Life
experiences affect each person differently by triggering different emotional responses and somatic markers. "The entire prefrontal region seems dedicated to categorizing contingencies in
the perspective of personal relevance" (Damasio 1994, 182). It
is critical to our ability to plan. Like a central information hub,
the prefrontal cortices are directly connected to every motor
and chemical response pathway that accesses the brain.

Note the parallel here with complex systems in general
(refer to Chapter 1). One of the hallmarks of complex systems is that variables affecting the system are often unseen.
Somatic markers are similar. They are able to act through our
consciousness as well as outside our consciousness. Emotional
responses of which we are not wholly aware often dictate
our responses to events. This is often referred to in colloquial
terms as going on "autopilot." These non-conscious emotional
responses are unseen variables that affect our decision making.

Intuition seems to be related to the somatic markers acting outside our conscious state. We may have a visceral, "gut
feeling" related to a situation where negative imagery gathered
from past learned experiences is generated that either inhibits
our tendency to act or encourages our withdrawal from the situation. Either way, its purpose is to greatly reduce our potential for negative decisions. The strong emotional reaction also
gains valuable time for us. Our intuition allows us to arrive at
a solution to the problem without having to reason through it
first. Creativity also comes from intuition. It is that wonderful
"ah ha" moment when the non-consciousness connects various lateral dots in the mind. Creativity is also influenced by
these non-conscious somatic markers. Damasio (1994) argues
that creative scientists have much in common with artists and
poets. Logical thinking and analysis are helpful to a scientist
when testing hypotheses but are not sufficient to facilitate

creativity. "Ah ha" moments or insights lead to breakthroughs but often not through preexisting conscious knowledge. It is the realm of the non-conscious, freely associating and making lateral connections among the incredible wealth of imagery representing experiences and knowledge in our mind. As Damasio quite accurately states, "creativity rest[s] on a 'merging of intuition and reason.'" (Damasio 1994, 189).

The decision-making mechanisms in the brain are body based and survival oriented, much like a squirrel racing to escape a predator. The mind rank orders the systems used, ranging from those associated with concrete decisions to abstract ones. The oldest system is involved in biological regulation. The next level deals with personal and social experiences, whereas the newest system in the brain deals with such abstract and symbolic operations as artistic and scientific reasoning, utilitarian or engineering reasoning, and the development of language and mathematics. All the stages are interdependent, as all complex systems are.

Although emotions are key to facilitating decision-making processes (Brooks 2011), they can also introduce irrationality.* Damasio suggests that within the human condition rationality often fails us in deference to a desire to obey, conform, preserve self-esteem, and so forth, which are often manifested by our emotions and feelings. He points out the example of people who are afraid of flying so they end up driving even though there is a greater likelihood of an accident while driving than flying. However, there is an aspect of a survival mechanism at play here. Planes do occasionally crash, and it is less likely that a person will survive that event than a car accident. We need emotions to make decisions and to make them in a timely manner; we need rationality to guide those decisions, minimizing the possibility of too great an emotional influence; in other words, we need to educate or train

---

* Ariely (2009) expands on and proves this point through compelling storied examples involving supposedly bright and rational MIT students and others.

our emotions (Brooks 2011). Examples include a pilot landing in bad weather and floor traders persisting in a volatile stock market exchange. Both are spurred on by the emotional drive of the situation, but neither is crippled by too great an emotional reaction.

Intuition helps us plan and decide when we are faced with a problem. Images of actions and outcomes come to mind, generating words and sentences about the images. Damasio uses an immunological analogy to describe how our intuition generates an incredible wealth of images related to possible actions we can take to overcome a problem. He refers to this as a "generator of diversity." To do this we must have a wealth of factual knowledge regarding the situation and possible actions and consequences. Our intuition categorizes the type of options, the type of outcomes, and the connections between the two, rank ordering the options in the outcomes to facilitate the decision-making process. Oftentimes we try to simplify our understanding of this process by referring to the common sense scenario: the brain detects a threat; it creates possible options; it chooses a response; it enacts the response; it handles the risk. However, the reality is much more complex:

> [N]eural and chemical aspects of the brain's response cause a profound change in which tissues and whole organ systems operate. The energy availability and the metabolic rate of the entire organism are altered, as is the readiness of the immune system; the overall biochemical profile of the organism fluctuates rapidly; the skeletal muscles that allow the movement of the head, trunk, and limbs contract; and signals about all these changes are relayed back to the brain, some via neural routes, some via chemical routes in the bloodstream, so that the evolving state of the body proper, which has modified continuously second after second, will affect the central nervous

system, neurally and chemically, at varied sites. The net result of having the brain detect danger (or any similarly exciting situation) is a profound departure from business as usual, both in restricted sectors of the organism ("local" changes) and in the organism as a whole ("global" changes). Most importantly, the changes occur in *both* brain and body proper. (Damasio 1994, 224)

In this complex system, the whole body and brain interact with the environment. A feedback loop is established as the body prepares itself be able to interact as well as possible with the environment. To maximize survival, the organism must continuously act on the environment and learn from those interactions. Learning is essential for the organism to survive because it enables the organism to be flexible and adaptable to the ever-changing environment. The neural interaction of the brain and the body creates what we know as the mind. An example of this is demonstrated with patients who suffer from spinal cord injuries. Even partially blocking the interactive signals between the brain and the body causes changes in the "mind state."

Synopsis

So what does this all mean in relationship to the thrust of this book? Perhaps the most important point to keep in mind when dealing with others, especially in complex systems engineering engagements, is that we are all human and we perceive things through different filters of personal experience, emotion, and thought. As stated in this chapter, this contributes to conflict in interactions with others due to our inability to understand each other's individual views of realities. Better understanding of human behavior and motivations, understanding ourselves and "the other," allows us to broaden our perspectives and hear others' views. This can lead to greater trust among individuals and greater willingness to collaborate in accomplishing common objectives.

Chapter 3

How to Build Trust

B. E. White and B. G. McCarter

There are two fundamental components regarding the object[*] of one's trust: (1) freedom from fear that the person, group, or organization will cause you, your group, or your organization physical, material, or psychological harm[†]; and (2) confidence in the veracity and accuracy of information provided by the person, group, or organization.[‡]

As suggested in the definition, two types of trust are important for information sharing: benevolence-based trust, where an individual, group, or organization will not intentionally harm another of the same type, and competence-based trust, where an individual, group, or organization is perceived to be

---

[*] In a general sense, notions of trust can be envisioned and interpreted as applying to groups or organizations, as well as individuals, or even to inanimate objects. Here we are emphasizing individuals or groups mostly, and then organizations, without any attention to inanimate objects.

[†] Psychological harm can be inflicted even in a relatively benign fashion; by being disloyal, for example.

[‡] More generally, you can judge the integrity of the object's observable actions. However, some of these, if not deemed directly relevant to your own interests, may not affect your trust much. Those that do are likely to fall into one of the above two categories.

knowledgeable* in some area (Levin et al. 2002, McCarter and White 2007).

Building trust between and among individuals and organizations is difficult, to say the least. Not only can achieving a high level of trust take a long time,† but trust can be lost very quickly and is difficult to regain. Trusting others is not a natural state for human beings. This is evident from the study of how humans have behaved—as individuals, groups, organizations, and even nations—through the ages. It is also a basic tenet of modern psychology. The culture of most organizations

---

* Knowledgeable means having the ability to act effectively (i.e., knowing how to perform a task well). This is not identical to being skillful, a quality associated with the actual performance of a task (i.e., with a dexterity or mental acuity better than most). One can be knowledgeable but not skillful or vice versa.

† However, there is the concept of "swift trust" (Iacono and Weisband 1997). "Sometimes there is no time to build a trusting relationship, such as when [a] group of people are thrown together and must start work immediately. A classic example of this is on the movie set. Make-up artists, key grips, stunt-men and many others are all on the job from day one, with little or no 'getting to know you' sessions. They must work out their differences on the fly and blindly trust one another to do their jobs. ... *Key factors that make for swift trust* [include] ... *Aligned activity ... Linked overall goals, rewards and penalties ... Interdependence ... Constrained environment ... Time ... Just-enough resources ... Non-person focus ... Professional role focus ... Task/process focus ... Trust broker ...* [who] *Hires, fires and leads the charge*" (http://changingminds.org/explanations/trust/swift_trust.htm). Jumpstart storytelling is one idea for facilitating the beginnings of trust in a hurry (Kahan 2006). Myerson, Weick, and Kramer (2006, 440) have expounded on swift trust as well: "what may be most distinctive about swift trust in temporary systems is that it is not so much personal form as it is a cognitive and action form. ... [S]wift trust is most likely when interdependence is kept modest through a combination of distancing, adaptability, resilience, interacting with roles rather than personalities, and viewing one's participation as partly voluntary (trust) and partly involuntary (confidence). In short, swift trust is less about relating than *doing*."

hinders information sharing* and often uses varying degrees of negative reinforcement (i.e., punishment) for doing so, because there is an absence of, or at least inadequate, trust. Over a lifetime, each of us probably has a wealth of first-hand empirical evidence to accept these statements as true.

On the other hand, we know that various types of trust can and do exist between and among individuals, groups, and organizations, and even nations, at least on a temporary (sometimes fleeting) basis. The evidence shows that trust can be achieved rapidly between or among individuals (even of differing ethnicities who experience nonthreatening face-to-face interactions on particular issues that gradually build momentum toward a shared intensity of common purpose (Carey 2008). However, trust can also be lost† rapidly, despite a track record testifying to a long and hard trust-building phase. So, how does the earning of trust come about? How can one engender more trust and enhance its persistence? Much comes from (1) behaving toward one another in a respectful and kindly way, and (2) the simple act of sharing useful or interesting information (that proves to

---

* "[I]ntellectual property is [an example of the tragedy of the] commons [http:// en.wikipedia.org/wiki/Tragedy_of_the_commons]. So how did Bill Gates become the richest man on earth? We are paying him rent. He privatized part of the 'general intellect', the social network of communication — it's a new enclosure of the commons. This has given a new boost to capitalism, but in the long term it will not work. It's out of control. Take a bottle of water: I produce it, you buy it. If I drink it, you cannot. Knowledge is exactly the opposite. If it freely circulates, it doesn't lose value; if anything, it gains value. The problem for companies is how to prevent the free circulation of knowledge. Sometime[s] they spend more money and time trying to prevent free copying than on developing products" (Else 2010).

† This could merely mean less trust or even that a state of distrust is reached. Trust can be thought of as a continuum, with no trust at one end of the scale and complete trust at the other; then distrust would refer to some contiguous segment including the no trust end, and trust to the complementary segment. Deciding where the breakpoint between distrust and trust occurs is moot.

be valid)—over time.* The literature is replete with suggestions on how to build trust. In this chapter we summarize some of that guidance, but more importantly, we synthesize what constitutes a trust-conducive state of affairs and suggest a way forward that will facilitate strengthening trust in organizations (perhaps the majority?) that have a strong chaordic nature, at least in terms of their informal underlying cultures. Finally, we opine what all this implies for leadership within decentralized organizations.

Looking ahead to later sections of this chapter, we anticipate leveraging traditional counseling psychology methods, which can help encourage trust and influence thought. Counseling Psychology examines human communication and facilitates individuals and groups moving through transformational change. It explores how we develop behavioral patterns through our interactions with others; how we communicate verbally and nonverbally; how that communication in a dynamic feedback loop can impact our interactions and relationships with others currently. All of this helps us understand ourselves and what in our experiential history has influenced our perspectives and behaviors.

In turn, this understanding helps individuals break the non-conscious automatic reaction to others that has been learned during previous life experiences, but which no longer work in the present situation and are maladaptive. Through counseling psychology and facilitative techniques, individuals learn that one's behaviors can be relearned and one's communication skills enhanced.

---

* We submit these two trust-building acts as axioms that do not require proof because they seem so obviously drawn from personal experience. It is unnecessary to delve deeply into scholarly research to find further evidence of their veracity. However, someone exhibiting these behaviors could have ulterior motives, and one should be somewhat wary of that possibility. As indicated previously, contrary events could turn one's trust into distrust; just because someone has behaved in a trustworthy fashion up to now does not imply that he or she always will (Taleb 2007).

Our views of reality, developed through past experiences, influence how we interpret and feel about events or other's behaviors; those feelings impact the way we, in turn, behave. Similarly, the way we behave can, in turn, impact the way we feel and the way we interpret current events or other's behavior. It is a dynamic feedback system.

When we are aware of the variables involved, we can begin to change our behavior and feelings, which can improve communication with others. This is often referred to as not being on autopilot. Being aware helps us break maladaptive behavioral patterns learned in the past that are no longer working in our interpersonal relationships today.

When we understand the many variables that have influenced the development of our view of reality (neurochemical, cultural, family experiences and values, sensory associations, and so forth), we can consciously engage in changed behaviors that modify those past communication patterns that no longer work effectively. Communication behavior changes can include tone, language, gestures, facial expressions, and how questions or comments are phrased. Use of these techniques can impact your communications with others, helping to build trust and collaborative behavior faster.

Perspectives on Trust

The rather cynical but interesting and informative book *The 48 Laws of Power* (Greene 2000) (refer to Appendix C of this chapter) provides practical insight into how people behave and how one might achieve greater (not so much control but) influence over them. The book is replete with mostly ancient and painstakingly detailed accounts of stories repeated to illustrate the various principles highlighted. A core idea for building trust is to try to get inside and be like the person or group you want to influence by minimizing their *us vs. them* reflex. (Refer to the NLP material above and the later subsection of this chapter entitled, Applying Laws of Power and Getting Inside.) An important part of this process is to expend serious efforts to: (1) discover their motivations; (2) understand their culture; (3) walk in their shoes; and (4) speak their language. It helps to be able to empathize (i.e., feel what someone else is feeling) (Pink 2005).

Fundamentally, one's ability to trust is strongly related to the negative emotions of suspicion, insecurity, and, let's face it—fear, a primeval human emotion. Unfortunately, trust and distrust are like an unbalanced coin that comes up tails (the negative side) too often.

We will discuss trust among individuals, or within and between groups and organizations. But first, as a backdrop, consider institutional trust (Burt 2001), a broad academic category considered to be a classic measure that ranges from insurance companies, investment firms and other financial institutions, online interactions (Pavlou, Tan, and Gefen 2003), and the media (Cook and Gronke 2001), to governments, countries, and nation-states. In particular, consider trust among nation-states and cultures, religious differences, and the role of fear in the world.

World Politics, Religions, and Fear

A fundamental principle of observing, studying, and working in complex environments is to think systematically and holistically, viewing the gestalt, or big picture. Considering our view of world history and how it has, does, and will affect us psychologically is an important perspective to have when thinking about issues of organizational and individual trust.

Arguably, the United States played a relatively minor role in the world prior to World War I (WWI), and even in the run-up to WWII compared to the post-WWII era. Since the Berlin Wall came down November 9, 1989 (Schmemann 2006), the United States has dominated world politics as the only superpower and has, in effect, taken on the role of the world's policeman, primarily by enjoying unprecedented and essentially unchallenged conventional military power. The duty of the United States to help engender world peace is no longer as clear as it was in WWII and its aftermath.* It can be argued that its leaders are not thinking as well as just subsequent to WWII when the United States created the (George C.) Marshall Plan, which enabled economic recovery and was perhaps instrumental in saving democracy in Western Europe and preventing the more complete domination of the Soviet Union in that region. Those nations that benefited from the Marshall Plan were eternally grateful to the United States. Since the Iron Curtain came down in 1989–90, that gratitude has ebbed gradually as the United States has increasingly exerted its power and, in contrast to its original behavior in the '40s and '50s, has tended to be less engaged with the United Nations in pursuing national interests.

Whereas before, the United States was rather small in influence, not much was demanded of it, and it had to prove itself and struggle to meet its needs; now it is very influential, much is expected of it, and it must redouble efforts to recapture its post-WWII position of high moral leadership. The challenge of communism manifested itself in several wars, starting with the Cold War, then Korea, and then Vietnam. Korea's stalemate and Vietnam's outlasting of the United States eroded its confidence as a nation and made it insecure again.

Uncertainty in complex systems or environments leads to polarization of groups within that environment. Polarization is characterized by a rigidity of thought or group identification that is very narrow, promoting extremism. Only the group's view of reality is tolerated. All other views are seen as a threat to the group. Often, those with opposing views are demonized and treated as *the other*. The current clash of cultures is

---

* "Nations must stop acting as though they can solve other nations' problems," wrote Ackoff (2004).

seen not only in the politics of the United States with the rise of various extreme positions and the demonizing of other groups, but also in the larger global conflict typified by the rise of Muslim extremism and the clash with the Western democracies.

The world is interconnected today through technology. That interconnectedness has exacerbated the global complexity. In turn, that complexity has increased the uncertainty and insecurity in the world. Polarization of groups and extremism are a logical result.

An understanding of complexity, uncertainty, and their impact on group dynamics whether at the local level or at the global level would help government policy makers address the fundamental issue of fear, while being mindful there are a host of other complex variables still impacting the actions of various groups in addition to fear. Actions that demonstrate this deeper understanding might contribute to rebuilding trust in countries like the United States.*

Complex human systems grow and thrive on diverse perspectives. Diversity drives creativity. And that creativity enables people to find and implement robust solutions to the serious issues the world faces today.

Political systems demonstrate these group phenomena well. As a result, psychology plays a strong role in U.S. politics. To the extent one has a firm belief in what one's own political party stands for or espouses, one is apt to misread or mischaracterize stated positions by members of an opposition party if those ideas resonate with the supposed platform of one's own party (Giles 2008). (This is part of the phenomena of group polarization and identification discussed in greater detail in Chapter 5.) Human fascination with rumors about prominent politicians and other important people also elicits debate

---

* According to Ackoff (2004, p. 1–2), "*One can only learn from mistakes*, by identifying and correcting them." and "it is better to do the right thing wrong than the wrong thing right."

about what to do about rumors. Recent psychological studies suggest that it is best to further publicize (false) rumors with rebuttals of why they are not true, as well as to shine a light on the individuals spreading the rumors and what their ulterior motivations might be (Singal 2008).

As Senge and others (2004) have pointed out, "the emergence of global institutions represents a dramatic shift in the conditions for life on the planet."* They are like a new "life form that, like any life form, has the potential to grow, learn, and evolve. But until that potential is activated, industrial-age institutions will continue to expand blindly, unaware of their part in a larger whole or of the consequences of their growth, like cells that have lost their social identity and reverted to undifferentiated growth for its own sake" (Senge et al. 2004, 4). (These global institutions are groups scaled up, and they demonstrate many of the same dynamics seen on smaller local scales.)

Why do people seem to assume that material and economic growth[†] are inherently good, and why do they equate economic growth with progress? Does it really increase happiness? Clearly, without finding innovative alternatives to the present shortage of critical resources, this growth cannot be sustained indefinitely (*New Scientist* 2008, 40) without diminishing the quality of human life on Earth, at least for

---

* Consider the work of Fukuyama (1992) with regard to nation-states versus religious states. Also relevant is Westphalian sovereignty, the concept of nation-state sovereignty based on two principles: territoriality and the exclusion of external actors from domestic authority structures. Many academics have asserted that the international system of states and the multinational corporations and organizations that exist today began in 1648 at the Peace of Westphalia (http://en.wikipedia.org/wiki/Peace_of_Westphalia). Both the basis and the result of this view have been attacked by revisionist academics and politicians alike, with revisionists questioning the significance of the peace and commentators and politicians attacking the Westphalian system of sovereign nation-states.

† We distinguish this type of growth from that of intellectual growth and maturity of thought in human beings, broader diversity in human interactions, and increasing complexity (e.g., growing and evolving states of innovation and integration) in human-made systems, all of which we deem as desirable forms of growth.

many. Critical natural resources are finite indeed, and the world's population continues to grow rapidly. Nature is all about checks and balances, growth and decay, life and death (Murray 2008, Senge 2008). Even if the world's population remained fixed, it seems likely that with our increasing rate of consumption, we will run out of sustaining energy resources before too many more generations go by. The independent nation-states, particularly those that are more industrially and technologically advanced, need to understand the inherent complex dynamics, trust more, and work together collectively to overcome the fears and narrow views of reality that hinder our ability to navigate the challenges our ever-changing world demands for the ultimate common good of mankind.

A good book (Kagan 2008) that primarily reviews the period since WWII offers a balanced perspective of where we are as nation-states and where we may be heading. Kagan aptly characterizes the permanent interests of the United States and how they have manifested in actions over these decades. These writings provide food for thought as a national overlay to individual beliefs and motivations that shape our organizations and workforces.

Biases of Individuals, Groups, and Organizations

A *New Scientist* article (Fisher 2006, 19) suggests that, based on anthropological studies, altruistic individuals may do well inter-group but selfish individuals do well intra-group. This is worthy of examination and discussion.* First, as already noted, trust is not the natural state of man. Assuming that as individuals we enjoy an identity and gain some affinity advantage by belonging to a group, it follows that we will cooperate with other members of our group to out-perform or beat out other groups in any competitive environment. Nevertheless, within our group, we are more likely to be competitive and pursue

---

* The origin of altruism is a controversial topic (Marshall 2010).

our own self-interest at the expense of other group members. If this is our goal, why should we share our information if this may provide help to other group members? Why should we trust our fellow group members? They are probably out to get us too! (For more on this topic, see the upcoming subsection, Perceptions of Reality and Power.)

In times of crisis, we might suspend our individual agenda for a time to support the group, but once the crisis has passed we typically revert to our old ways. One of the mysteries, which the Enneagram™ group process (see Chapters 4 and 5) that Richard Knowles advocates (Knowles 2002) is meant to alleviate, is why we cannot behave in a more benevolent fashion (i.e., embracing trust as our normal behavior). However, maybe the best we might hope for is to achieve more trustworthy behaviors as a group. It may never be possible to sustain trustworthy behaviors among all the individuals with whom we interact. There may be just too many perceived threats (fears), individual personality issues, and temptations to expect that.

Learning in Organizations

Individual learning, especially as established as schoolhouse approaches by institutions, can work against organizational learning because connections between emotional dynamics and organizational politics may be ignored, or at least not addressed. As discussed to some extent in Chapter 2 and Chapter 4, emotion can be a key component of organizational learning. As individuals learn and assert their newfound knowledge, some managers tend to be challenged and feel threatened by their subordinates. These managers become defensive and push back; this is not conducive to their own (or organizational) learning, in general. Exhibiting emotion (e.g., crying) in the workplace is frowned upon because it makes others feel uncomfortable.* In some organizations this can even derail promising careers. It takes a considerable level of individual and group trust to allow emotions to be acknowledged as a positive influence in organizational learning. Suppressing the acknowledgment of people's feelings can be counterproductive to the long-term effectiveness of an organization. Even high performers who keep company interests their top priority can be criticized and lose their favored status if they become too emotional and overreach and fail, or burn out through overwork (see, e.g., the hypothetical case of Stephen of the Goodwill Company, Vince 2002).

Argyris (1991) speaks of the more fundamental "double-loop learning" and how successful people tend to blame others for problems they ultimately face as they rise in an organization, rather than realizing and acting on the need for their own learning. Another problem (Argyris 1994) is executives who create unrealistic visions of their company as a happy place for their employees and who promise to do things for them. They defeat organizational learning because the troops are not challenged or motivated to accept responsibility for organizational success through their own collective action. Individuals learn best when they are challenged, but not when they are challenged to the point of being overwhelmed. It is an important distinction to make.

A learning organization can be defined as a place where "employees continually create, acquire, and transfer knowledge" (Garvin, Edmondson, and Gino 2008, 109). Organizations need to build in learning time if they want

---

* Instances of political candidates crying in public have had a similar detrimental effect. For example, Edmund Muskie became the favorite to win the 1972 democratic presidential nomination. But being the front-runner for over a year proved difficult. During the New Hampshire primary, Muskie choked with anger and seemed to cry because of a couple of nasty articles in the *Manchester Union Leader*. One article proved to be a hoax. The other attacked Muskie's wife. Muskie then attacked publisher William Loeb. The episode came to symbolize the collapse of Muskie's presidential campaign because of the perception that he was weak (http://www.pbs.org/newshour/bb/remember/muskie_3-26.html).

to achieve the ability to more effectively deal with unexpected events and survive over the long term (Edmondson 2008). From that perspective, devoting all resources to short-term needs and pleasing your customers only through efficient execution of the work at hand can be short sighted. To achieve the goal of becoming a learning organization, top and middle management need to create an environment where employees feel psychologically safe in spending some of their time learning. At the same time customers need to be made to understand that the quality of the services provided by the organization will be enhanced if some resources are allocated to acquiring knowledge and learning skills related to forthcoming or eventual products.*

The Harvard Business School has surveyed executives to establish a database from which to evaluate the relative degree to which any organization that completes the survey can be considered to be a learning organization. The survey questions are binned into three categories related to what are considered building blocks of learning organizations.

1. A supportive learning environment[†]
   - Psychological safety
   - Appreciation of differences
   - Openness to new ideas
   - Time for reflection
2. Concrete learning processes and practices
   - Experimentation
   - Informal collection
   - Analysis
3. Leadership that reinforces training
   - Education and training
   - Information transfer

---

* Unfortunately, too often customers are impatient in their desire for short-term results and resist learning investments for the longer term; further, management of the service organization fails to make the case for the latter.

† Trust is particularly relevant to this first block.

The questions are available in the article (Garvin et al. 2008) and can be accessed online (https://surveys.hbs.edu/perseus/se.ashx?s=381B5FE533C282FF). After looking over the questions, consider how groups of employees within your organization might answer them. Then imagine how their management might feel about the results. In many cases, management would probably not like the answers, at least for organizations that think of themselves as learning organizations but do not consciously devote attention and resources to building these blocks.

Storytelling

Not surprisingly, a good amount of interpersonal trust can be built by storytelling. Think about your work friends and the opportunities you have to interact on a more personal level outside of work hours. What do you do a lot of? Storytelling! Think of the power of storytelling for getting your message across. Why can't we incorporate more of this during our regular workday?

Kahan (2006) advocates what he calls "jumpstart storytelling" to accelerate interpersonal collaboration. This technique can be applied effectively at the beginning of a meeting where many people do not know each other well:

1. Divide the attendees into groups of three to five people.
   - Select a general topic appropriate for the meeting and explain that choice to the assemblage.
   - Ask that each person in each group tell a personal story (of no more than ninety seconds) about that topic to other members of their group. After only a matter of minutes, everyone has heard each other's stories.
2. Shuffle the groups and have everyone repeat their story to their new group.
   - This should take no more than about fifteen minutes.
   - Ask everyone to move to the person whose story they liked the best and put a hand on that person's shoulder.

– Those with the most hands on their shoulders are then asked to tell their stories to the entire group.
– Discussing why these popular stories are compelling can lead to insights about the topic and be a springboard for further discussion at the meeting.

More important, perhaps, is how the participants have started building interpersonal trust through their storytelling connections and will be more open to collaborating on the topic during and subsequent to the meeting. Try this technique at one of your future meetings, and see what you think.

Storytelling enables us to make sense of complex events or dynamics. It allows us to see possible alternative results for decisions that can be made.

The underlying psychological dynamics of storytelling that reflect the living system of human dynamics include the following:

Identity:
– Allows us to try on the roles of others, experience their perspectives
– Helps us see issues from different perspectives, make decisions in a different environment
– Influences our sense of self
– Shapes our emotions and our actions
– Enhances or changes our perceptions
– Involves our feelings and emotions

Relationships:
– Develop rules of behavior
– Develop community through a shared story
– Influence our group identifications
– Impact our interactions with others

Information:
– Allows us to see consequences of our actions
– Helps us understand complex dynamics
– Changes the way we think

Excellent pointers for effective storytelling (Denning 2005) and how to make your ideas stick (Heath and Heath 2005) are worth reviewing, internalizing, and building into your daily life. Increasing others' trust in you is an important by-product. Denning's storytelling rules can be summarized as follows:

- Start from where the audience is, not from where you are
- Present an unexpected, relevant, but negative situation to get audience's attention
- Tell how things will get worse unless effective action is taken
- Stimulate a positive desire to achieve a happy ending
- Speak the truth, or at least the plausible; be memorable
- There are two listeners (the one you see and the little voice in their head)
- The little voice generates a springboard story from the impact of the story you have told

The Heath brothers (Heath and Heath 2005) employ the SUCCESs acronym as an aid to memorizing how to present your ideas so that others will understand, retain, and (one hopes) adopt them:

- **S**imple (compact and addresses one of your core issues)
- **U**nexpected (arouses your curiosity to know more)
- **C**oncrete (expressed in real terms to which you can relate)
- **C**redible (authoritative and passes your giggle test)
- **E**motional (grabs you in the gut)
- **S**toried (you can imagine how it affects your world)
- People will remember **SUCCESSFUL** ideas!

Note the overlaps and similarities in all of these trust-building techniques.

Perceptions of Reality and Power

Most humans are subject to the attribute of rigidity of thought to some degree. We all strive toward a unified but elusive perception of reality to reinforce our own perceptions, an evolving collection of sensory, cognitive, or experienced, processed, and internalized lifetime happenings. We are continually trying to make sense of the world. In others, we try to find alignment or validation of our beliefs. Rigidity of thought includes the rejection, discounting, or rationalizing away of others' beliefs or behaviors that do not mesh with our own.

As previously noted, we fundamentally do not trust others because of fear. We tell ourselves a story, consciously, or even unconsciously, to reinforce what we do know and understand, and to fill in the blanks of what we do not know, in ways that are compatible with our currently perceived notions. Because we are all fallible due to our many filters through which we perceive our world, it is likely that none of us sees the true underlying reality of anything. We tend to join with others that happen to share our (often flawed) perceptions in order to build our own confidence that we are on the right path through life, reinforcing our sense of self and the reality we have constructed through our experiences. As we all know intellectually, this has its dangers (e.g., groupthink), as even John F. Kennedy realized—after the fact—when he consulted with his closest advisors during the Bay of Pigs fiasco (http://en.wikipedia.org/wiki/Bay_of_Pigs_Invasion, http://www.jfklibrary.org/Historical+Resources/JFK+in+History/JFK+and+the+Bay+of+Pigs.htm, http://www.u-s-history.com/pages/h1765.html).

So rigidity of thought is driven by insecurity, taken to its ultimate—the fear of death. But in the immediate reality, our basic fear is not being able to handle a world we do not fully understand.

Affinity with groups to which one belongs and the desire to enlist the support of and behave like the group contribute to the group's ability to have a more cohesive purpose and to influence other groups of like mind, or if not so much of like mind at least part of the same community of interest. The characteristic remains, however, that the ability to trust is easier for a group as a whole with respect to interactions with other groups than it is for individuals with respect to interactions with other individuals in the same group. Most likely, this is influenced by the group's reasoned, debated, and collective view being typically seen as more correct than that of an individual member.

Nevertheless, there are good reasons to not trust people. On the basis of considerable research, Stout (2005) points out that approximately four percent of the human population has no conscience! In other words, on the average, one person in a group of twenty-five may choose to pursue a particular course of action to their own perceived advantage and feel absolutely no guilt, despite the potential negative, and sometimes devastating, impacts on others. Of course, there may not be such a culprit in your specific group of twenty-five, especially when the group represents a relatively homogeneous set of people from the same organization. As you might envision, however, this statistic can be used in a humorous vein during a meeting to gently suggest that someone in the room might have ulterior motives. Stout also explains how identifying such individuals is difficult because conscience-free manipulators are so good at coming across as likeable human beings to their constituents, at least on the surface, which is often good enough for these operators to get what they want. If and when discovered, and if not prosecuted because of lack of evidence or provable plausibility, such human pariahs move on to other venues and begin again.

Applying Laws of Power and Getting Inside

Refer to Appendix C: Quotations from *The 48 Laws of Power* (Greene 2000), from the book mentioned at the beginning of this section, for a hefty dose of Machiavellian advice from a very cynical but perhaps instructive or illuminating point of view. The examples highlighted there may be viewed as the behaviors of people with no conscience or perhaps the behaviors that might help combat such people.

These behaviors are the antithesis of what NLP and the Process Enneagram™ facilitation technique strive to teach. The behaviors discussed here represent a perspective of control and domination over others that, instead of consolidating power, actually encourages passive-aggressive resistance to mission goals because buy-in and true consensus are never achieved. When people do not feel their voice or opinion is heard, they will push back in non-assertive ways. Nevertheless, the quotations embody precepts of which we should become more aware.

In organizations with hierarchical management structures, power and monetary reward are generally thought to be in direct proportion to one's level in the hierarchy, although this is not always true.* The assumption about rewards motivates many people to aspire to promotion. Should one just trust the system to offer promotions in a just and equitable fashion? Not necessarily.

An individual's perception of personal power is partly built on a shared perception of reality that others acknowledge as being valid. This reinforces one's sense of well-being. Some people wish to increase their personal power and seek to do so by convincing others, perhaps in more subtle than overt ways, that they should have it. For these power seekers, if others perceive that they should have power, so much the better! That saves them from having to mimic those that have no conscience, for example, in order to get power.

If so desired, how do you orchestrate the accrual of power by acclamation from superiors (especially) or peers? Perhaps eliciting this mandate from subordinates is easier? Not if your subordinates are trying to advance in this area as well, and even surpass you. This is trickier than overtly convincing people that you should have power. You may have to operate

---

* Certain talented individuals in non-managerial staff positions who create significant value for the organization often are rewarded handsomely and enjoy their ability to have influence.

in ways or venues that are beyond your present hierarchical status to exhibit your capabilities. However, even if you do well in this, you may still have to ask for the promotion before management thinks of offering it. There are dangers in being too obvious about this. Those above you may feel threatened, and peers or subordinates might wish to sabotage your chances. So this overt behavior could set you up for failure. Understandably, setbacks or recriminations can increase your own insecurity and level of fear and, possibly, in more extreme cases, even lead to lashing out at others in retaliation. Not good. On the other hand, many promotions are viewed with little surprise because the individual being elevated is generally recognized as, in effect, performing well at the new level already. Nevertheless, there is often a higher level mentor behind the candidate who brings the candidate along.

Another way to choose leaders is almost by acclamation from co-workers or associates acting from the bottom up because of deep-seated (ancestral) preferences for people of adeptness in a certain domain and/or of apparent stature (tall, square jaw, etc.) and affiliation with the group (being one of us), for example (Ahuja 2010). This is less likely to occur in hierarchical organizations because traditional power is so entrenched with upper management. Decentralized organizations that have a chance to act in a more democratic fashion may be able to follow this approach. Then again, to the extent this bottom-up approach is successful, more hierarchical organizations may experiment with its adoption.

Types of Trust

Persistent Trust

Recall Aesop's fable about the tortoise and the hare: Those who persist often succeed in achieving their goals, almost no matter what the setbacks. Having the philosophy that failure comes only from giving up seems to be a trait of many ultimately successful people. Thomas Edison, credited with being a genius,* is thought to have pursued a hundred technical ideas for every one that proved practical. One kind of trust is associated with the confidence that an individual will keep plugging away, continually trying new things or alternative paths to their destination. Most people would like those types to be on their teams, and managers or leaders usually welcome them as well.

Many bosses would value a certain degree of arrogance and self-confidence in their employees because this can help them be persistent in the pursuit of goals (Phillips 2006, 17).

> Self-belief can make for better managers and decision makers because self-assured people feel able to take the initiative, make choices [on their own], … in order to get things done. Another benefit of strong beliefs is that they can make people appear more authoritative—and that makes other people believe what they are saying.

Sometimes, however, persistence may have negative consequences; for example, when someone is trying to say no without saying it. You may have a pet peeve about people (especially those who are the single addressee) who do not respond to e-mail, even after more than one attempt. Some recipients share a culture of never saying an overt "no" to internal e-mail requests for help (e.g., funding support). Rather, the request is simply ignored until it goes away. The thought is that requesters will understand this and not pursue the matter further. Even though the requesters may get

---

* Indeed, this is difficult to refute, although Edison probably benefited greatly from other researchers, such as Nikola Tesla, whom he did not encourage nor give much credit to (http://en.wikipedia.org/wiki/Thomas_Edison, http://en.wikipedia.org/wiki/Nikola_Tesla).

the message being conveyed, they may not like it and thus persist in pursuing a direct answer, next time with a telephone call or personal visit. Another culture would be that everyone deserves an answer to such e-mail requests; hopefully, a straightforward and honest one, but at least one that contains some kind of rationale. For example: "Sorry, we cannot afford that within our current budget and spending projections, at least without adjusting our priorities, which we do not feel is warranted."

The non-response is an example of a common power tactic that typical senders would likely not do much about. One response that sometimes helps is to persist in continually raising the issue until those reluctant to respond finally do. It tends to catch them in their behavior, and if they are embarrassed they may feel compelled to respond next time. Perhaps more likely, however, this might reinforce their initial act of ignoring you, and they will become even more passive-aggressive in future interactions. In this instance, persistence can lead to a negative result.

In this complex world, it is rarely easy to make significant progress on difficult problems quickly, unless you are extremely lucky. Of course, you make your own luck, and persistence is a key element in doing that.

Contemplating action often conjures up elements of risk. Many people are reluctant to act because it can be risky. Those who are risk-averse ask themselves, *What if I fail?* and tend to not pursue the next step unless they feel the chances of success are likely. Certainly it is prudent to try to mitigate downside risk (Taleb 2007). Some advocate the adage: Always think of the worst thing that could happen before deciding to embark on a course of action. The implication is to do nothing if some negative outcome, if incurred, would be too painful. Not much would get done in this world if everyone followed that philosophy.

In complex environments of enterprises (Rebovich and White 2011), extended enterprises, mega-systems (Stevens

2011), and so forth, one should be more attuned to opportunities than to risk (White 2006) because things are so dynamic and out of one's control that there's little hope of even predicting, let alone guaranteeing, anything. The biggest risk in these environments is to *not* pursue opportunities because that could eliminate many possibilities for improvement and perhaps even unforeseen solutions. In such situations, persistence can be similar to playing chess, continually trying to improve your position in the face of opposition moves, and never feeling one has lost as long as there is some hope of winning.

*How Your Emotions May Affect Your Trust**

It has been shown that our incidental emotions affect trust more with acquaintances than with people we know well. Our emotions can have a greater effect on judgment when they are not salient, as opposed to being highlighted and consciously acknowledged. Here is a sampling of what Dunn and Schweitzer (2005) found:

- Happy, sad, and angry participants have significantly decreasing levels of trust. (p. 740)
- Gratitude and anger of others influence trust more than one's pride and guilt. (p. 741)
- Grateful participants are significantly more trusting than others. (p. 742)
- Incidental emotions could change the way others judge *their* trustworthiness. (p. 746)

---

* Refer to Chapter 4 concerning how emotional intelligence (Goleman, Boyatzis, and McKee 2002) can affect leadership in groups.

■ Individuals should curtail the influence of incidental emotions on their judgment through increasing awareness of the sources of their own and others' emotions but this is difficult because of personal biases and differing social perceptions. (p. 746)

In any event, it is important to recognize that emotion plays a significant part in one's ability to trust. Descartes was wrong in saying that humans are purely rational beings (Damasio 1994).

*Interpersonal Trust When Not Face to Face**

Issues surrounding interpersonal trust are better understood in situations where individuals have relatively frequent face-to-face contact. In this section we explore what is becoming the more common situation in many organizational and online learning[†] environments, where we do not have significant face time with the people with whom we are expected to interact. This is particularly true in chaordic organizations, where the burden of travel costs decrease face-to-face meeting opportunities (Abrams 2003). In these situations people tend to interact in other ways using multimedia modes of communication; these may include e-mail (primarily), telephone, cell phone, iPad or tablet, as well as other means of audio or video teleconferencing. Increasingly, people are also exploring virtual networks (refer to Appendix D of this chapter), such as Second

---

[*] Refer to Chapter 4 concerning leadership traits that seem to best engender trust in virtual teams (Pierce and Hansen 2008).

[†] "Establishing trust quickly is the key to effective Internet communication, especially when it comes to teaching online. ... The most effective online teachers establish communication early and quickly." They do this by quickly establishing a climate for warmth and responsiveness. "By 'swiftly' replying to each student's initial comments, ... the most effective professors provide students with a sense that there really is a professor at the other end of the communication link" (*ScienceDaily* 2005).

Life (http://secondlife.com),* MySpace (http://www.myspace.com), LinkedIn (http://www.linkedin.com), and Facebook (http://www.facebook.com), even using avatars of themselves. We intend and think it is important to begin comparing and contrasting how people gain, maintain, and lose trust in these chaordic situations versus face-to-face interactions.†

Let us first consider the latter situation outlined above, when you might interact with others in a virtual way using an avatar to represent your persona. This may make you feel less vulnerable and thus encourage you to behave in a manner inconsistent with how you would normally behave in face-to-face encounters. The avatar may provide a feeling of increased security because the real you is hiding behind a mask. In effect, you may be disguising your true feelings. This can make it more difficult for others to trust you, especially to the extent your network behavior becomes inconsistent.

However, this mode can help you more effectively probe others to gather information about how they behave‡ and allow you to test the consistency of their responses in a more anonymous fashion. To some degree, this can become a game with more variation than we find in face-to-face encounters.

Alternatively, as you interact with other avatars, you may actually become more vulnerable if you ascribe too much credibility to and naively believe what you observe (i.e., in effect, you can trust too much). A humorous and ultimately happy example of this is exemplified by the movie *You've Got*

---

* For a recent (mixed) review of Second Life in terms of virtual reality and serious games, refer to Julian Dibbell's short article in *Technology Review*, January/February 2011, 74–76. Also, for an excellent introduction to serious games and the impact on learning and building of trust in non-face-to-face situations, refer to http://www.ibm.com/ibm/files/L668029W94664H98/ibm_gio_gaming_report.pdf and http://www.vizworld.com/2010/01/ibm-asks-virtual-worlds-real-leaders/

† Certainly, there is anecdotal evidence that high levels of trust can be engendered among dedicated teams working on important projects whose members never see each other (William D. Miller, personal communication, October 17, 2008).

‡ This could be viewed as a mild form of Machiavellian behavior on your part. For more, see Appendix C.

Mail, starring Tom Hanks and Meg Ryan. A regrettable such incident surfaced in the news in June 2008 (Cathcart 2008). Over the Internet, an adult woman convinced a thirteen-year-old girl that she was communicating with a boy while she abused the girl with messages. The girl became so distraught over "his" attacks that she committed suicide.

So far, the research is limited and divided as to whether virtual communication through avatars or anonymous online personae increases or decreases the ability to trust. We would like to see more exploration of both sides of this question. For example, oxytocin is an endorphin chemical of the brain that is produced in greater quantity with face-to-face contact of the kind that engenders trust (e.g., lovemaking) (Fisher 2005). One question is the degree to which oxytocin is generated without face-to-face contact. Current research (refer to Appendix D in this chapter) may lead one to think the same effects are indeed achieved through 3D immersive virtual environments.

Trust and Inter-Reality Systems

Another fascinating area of trust is mixed or dual reality states, in which subjects are immersed in situations that are partly real and partly simulated or artificial (Hubler and Gintautas 2008). Individuals in these mixed reality states, where they and the virtual system are in synchronism, must trust the situation. Some experimental evidence suggests that sometimes they trust too much, even mistaking the virtual system for themselves (Thomson 2008).

Hubler and Gintautas (2008) discuss examples of mixed and dual reality states. The following one is pretty straightforward. When you drive a car, you have a mental (virtual) image of you and the car (the reality) moving down the road. When the car responds to your actions as expected, you enter a mixed reality state. If the car does not respond appropriately (e.g., if you lose control of the car), you enter a dual reality state. Then the virtual and real states are not in synchronism as they would be in the mixed reality state. In another aspect of mixed reality, your mental image can be advanced in time (e.g., you may be able to imagine how you will attempt to control the car down the road in a patch of arising trouble).

Here is another example that can be a little harder to imagine unless you have experienced it firsthand. Suppose a subject is asked to sit in a chair facing a screen where her image from a camera that is placed arrears but forward-facing is projected. When someone standing behind the subject places a hand on her shoulder, she has the sensation, while watching the image, that she is seeing herself being so touched from outside her body. She is in a mixed reality state. Prof. Hubler demonstrated this live with a voluntary subject during his talk at the University of Illinois in May 2008 (Hubler and Gintautas 2008). When asked by Dr. Hubler where she was, the subject slowly raised her hand and pointed at the monitor, saying she was there.

Trust in Mental Health

Immersive virtual environments are taking an increasing role in medicine and psychology in treating patients who are mentally ill or recovering from brain trauma (Rizzo 2008). Patients are exposed to virtual images of other patients (avatars) that the therapist can manipulate to show positive or negative (i.e., rational and irrational) behaviors to which the real patient can relate. Apparently, this is being contemplated as an exploratory form of therapy based partly on the premise that the patient may be more willing to trust the avatar than the therapist. Presumably, the patient will tend to mimic the behavior of the avatar, and the therapist can explore ways to gradually bring the patient out of depression or whatever dysfunctional state they may be in.

Other compelling examples include mental health therapy using toy (as well as live) animals with elderly bed-ridden patients and with others experiencing recent bereavement, for example (Fine 2006). Patients are able to build greater trust and contentment as a result of feeling responsible for their charges.

In addition, virtual simulations are being used to treat soldiers suffering from psychological disorders such as post-traumatic stress disorder (PTSD) (Magnuson 2008). These techniques are becoming part of the mainstream.

Trusting with Limited Interpersonal Contact

Intermediate situations are those in which there is limited face-to-face contact but considerable interaction via e-mail in a networked environment of similar interests. Anecdotally, at least, considerable trust can be built over time prior to, between, or subsequent to face-to-face meetings.* The second author of this chapter (B. G. McCarter) experienced this within a network of mothers of profoundly gifted children. Even though she met others face to face only once or twice, trust was built continually using e-mail to share personal information, tell stories, compare notes, and express empathy (Pink 2005). So it appears that even in these more impersonal environments, the same trust-building principles as for face-to-face contact can apply.

By way of contrast, dating websites (e.g., http://www.match.com/matchus/, http://www.millionairematch.com/) in themselves do not engender trust; they are more of a filtering mechanism. The trust building occurs primarily through meeting personally.

Trust is almost automatic with people you have known for a while and with those to whom you are reasonably close. When encountering them even after a long absence of face-to-face contact, you tend to pick right up where you left off. It may take a number of negative incidents for you

---

* The (now-deceased) spouse of the first author of this chapter (B. E. White), an accomplished salesperson and sales executive, often gave the following advice to her colleagues: "Your goal is to make the potential client your friend. He or she is more likely to buy our product if they trust you personally."

to lose that level of trust (Kleiner 2005). Trust is the utility by which the considerable store of tacit knowledge flows. In organizations with trust, more value is created and wasteful overhead is reduced. The competitiveness of organizations is much more associated with the quality of their informal human networks than has been thought, according to Prof. Karen Stephenson, a guru in organizational development and change, with very impressive credentials in making good things happen, despite an early background in fine arts, anthropology, and chemistry. Stephenson does not advocate the abolishment of hierarchy in organizations. Rather, she believes in a hybrid organization that blends hierarchy and information networks, something the authors of this book support as well (McCarter and White 2007). She highlights the importance of people who, in effect, serve as hubs to the shortest paths to information useful to the organization (i.e., the connectors). Prof. Stephenson did her PhD dissertation on a study of Bolt, Beranek, and Newman (BBN). Bolt was a personal friend of Margaret Mead. He said if anyone can understand BBN, it's an anthropologist! Stephenson said the flip side of trust is betrayal.

Here is a final thought about trust with limited interpersonal contact. Although distributed networks can be used to make significant inroads in accomplishing big organizational goals, it is unrealistic to expect them to always reach a tipping point (Gladwell 2002) of fundamental phase change (Robertson 2003). Suppose you are trying to get your organization to do something really significant (e.g., help change the way the world works in your particular domain, over a long period of time, such as decades). The typical reactions you may receive from those in your network are: "You're tilting at windmills!" (like Don Quixote) or "This is too hard. There's no way you can change the system. Let's not even try!" Can you trust your networked colleagues to listen to your ideas and perhaps embrace them if they seem worthwhile, and maybe spread them around to garner additional support? Or can you trust them to at least root for you and provide some encouragement? To help combat the frustrations that might accrue with negative answers to these questions, it is important to maintain perspective. Realize that fundamental change often takes a long time, especially when the culture of an organization is so deeply ingrained that such changes will be strongly resisted by both the rank-and-file and senior management. It may take a couple of generations of new leaders for transformational ideas to percolate sufficiently through an organization to instigate significant change.

As we have discussed, human emotion is important in establishing trust (or not). Individuals are more likely to view change, particularly organizational change, as a positive thing if they view it through a lens of opportunity growth rather than career-ending risk. This is a notion that good leaders and mangers espouse and understand.

Implications for Leadership

It is incumbent on organizational leaders to understand how trust works among their constituents. This is particularly true in decentralized organizations, where leadership and management challenges tend to be greater because of the fewer face-to-face opportunities available (Abrams et al. 2003). Abrams' article provides evidence that among distributed organizational networks that discover personal expertise on topics of strategic interest, individual contacts are much more effective than an expensive knowledge-sharing technological infrastructure.* The two fundamental traits in others that engender trust are benevolence and competence.

Quoting from Abrams:

---

* However, this is not to discourage outstanding knowledge management efforts within organizations. Good knowledge management has its role and degree of usefulness, albeit the utility needs to be measured, if possible, and balanced against the cost.

From our interviews, we learned that those who are seen as trustworthy sources of knowledge tend to: (1) act with discretion; (2) be consistent between word and deed; (3) ensure frequent and rich communication; (4) engage in collaborative communication; and (5) ensure that decisions are fair and transparent. (Abrams et al, 2003, p. 65).

… Under organizational factors, we identified two ways to promote interpersonal trust: (6) establish and ensure shared vision and language; and (7) hold people accountable for trust. … Under relational factors, there is some overlap with the trustworthy behaviors mentioned above, but we also identified two new behaviors: (8) create personal connections; and (9) give away something of value. Finally, under individual factors, a person's own judgment of his or her abilities … [matters] which we characterize as (10) disclose your expertise and limitations. … (Abrams et al, 2003, p. 65- 66)

Dunn and Schweitzer point out an import caveat that should be noted in addition to the points made by Abrams. They note that decision makers should avoid making quick trust decisions and, instead, take precautions to make trust judgments over time and on the basis of interactions across multiple contexts. (Dunn and Schweitzer 2005, 746)

Table 3.1 is an apt summary of things a manager can do.

Recapitulation

Here we review the many perspectives on trust and types of trust discussed in this chapter in the context of what a forward-looking organizational leader might do. These are suggestions and by no means should be viewed as complete. We

Table 3.1   Managerial Behaviors That Promote Interpersonal Trust

Trust Builder	Description and Logic	Managerial Actions
Trustworthy Behaviors		
1. Act with discretion	Keeping a secret means not exposing another person's vulnerability; thus, divulging a confidence makes a person seem malevolent and/or unprofessional. *Promotes: benevolence trust*	Be clear about what information you are expected to keep confidential. Don't reveal information you have said you would not … and hold others accountable for this.
2. Be consistent between word and deed	When people do not say one thing and do another, they are perceived as both caring about others (i.e., they do not mislead) and as being competent enough to follow through. *Promotes: benevolence and competence trust*	Be clear about what you have committed to do, so there is no misunderstanding. Set realistic expectations when committing to do something, and then deliver.
3. Ensure frequent and rich communication	Frequent, close interactions typically lead to positive feelings of caring about each other and better understandings of each other's expertise. *Promotes: benevolence and competence trust*	Make interactions meaningful and memorable. Consider having some face-to-face (or at least telephone) contact. Develop close relationships.

| 4. Engage in collaborative communication | People are more willing to trust someone who shows a willingness to listen and share; i.e., to get involved and talk things through. In contrast, people are wary of someone who seems closed and will only answer clearcut questions or discuss complete solutions.

Promotes: benevolence and competence trust | Avoid being overly critical or judgmental of ideas still in their infancy.

Don't always demand complete solutions from people trying to solve a problem.

Be willing to work with people to improve jointly on their partial ideas. |
| 5. Ensure that decisions are fair and transparent | People take their cues from the larger environment. As a result, there is a "trickle down" effect for trust, where the way management treats people leads to a situation where employees treat one another similarly. Thus, fair and transparent decisions on personnel matters translate into a more trusting environment among everyone.

Promotes: benevolence trust | Make sure that people know how and why personnel rules are applied and that the rules are applied equally.

Make promotion and rewards criteria clear-cut, so people don't waste time developing a hidden agenda (or trying to decode everyone else's) |

(continued)

Table 3.1 Managerial Behaviors That Promote Interpersonal Trust (Continued)

Trust Builder	Description and Logic	Managerial Actions
Organizational Factors		
6. Establish and ensure shared vision and language	People who have similar goals and who think alike find it easier to form a closer bond and to understand one another's communications and expertise. Promotes: *benevolence and competence trust*	Set common goals early on. Look for opportunities to create common terminology and ways of thinking. Be on the lookout for misunderstandings due to differences in jargon or thought processes.
7. Hold people accountable for trust	To make trustworthy behavior become "how we do things here," managers need to measure and reward it. Even if the measures are subjective, evaluating people's trustworthiness sends a strong signal to everyone that trust is critical. Promotes: *benevolence and competence trust*	Explicitly include measures of trustworthiness in performance evaluations. Resist the urge to reward high performers who are not trustworthy. Keep publicizing key values such as trust—highlighting both rewarded good examples and punished violations—in multiple forums.

Relational Factors

8. Create personal connections	When two people share information about their personal lives, especially about similarities, then a stronger bond and greater trust develop. Network connections make a person seem more "real" and human, and thus more trustworthy. *Promotes: benevolence trust*	Create a "human connection" with someone based on nonwork things you have in common. Maintain a quality connection when you do occasionally run into acquaintances, including discussing nonwork topics. Don't divulge personal information shared in confidence.
9. Give away something of value	Giving trust and good faith to someone makes that person want to be trusting loyal, and generous in return. *Promotes: benevolence trust*	When appropriate, take risks in sharing your expertise with people. Be willing to offer others your personal network of contacts when appropriate.

Individual Factors

10. Disclose your expertise and limitations	Being candid about your limitations gives people confidence that they can trust what you say are your strengths. If you claim to know everything, then no one is sure when to believe you. *Promotes: competence trust*	Make clear both what you do and don't know. Admit it when you don't know something rather than posture to avoid embarrassment. Defer to people who know more than you do about a topic.

Source: From Abrams, Lisa C. 2003. "Nurturing Interpersonal Trust in Knowledge-Sharing Networks." *Academy of Management Executive* 17(4):(67) 64–77. With permission.

encourage you to treat this as a checklist to be embellished and expanded on while leveraging your own creative juices.

- Recognize the importance of human psychology and human emotion and be more humble in your own beliefs.
  - Don't automatically assume that you are correct in all things.
  - Think of others and the Golden Rule.
  - Focus on quality, not growth for its own sake.
- Lead in creating an environment for group communication.
  - Be comprehensive in addressing all aspects of each topic.
  - Create an open environment for expressing diverse perspectives.
  - Apply the Process Enneagram™ for self-organizational facilitation.
- Create conditions for self-learning.
  - Motivate by designing facilities and mechanisms that are emotionally compelling.
  - Express challenges and possible approaches to progress that help build attitudes of self-sufficiency and confidence.
- Help others understand your ideas through storytelling.
  - Build storytelling into the everyday life of the organization.
  - Learn, exemplify, and spread the principles of making ideas stick.
- Guard against rigidity of thought in yourself and others.
  - Keep minds open and questioning.
  - Avoid groupthink.
  - Be sensitive to and realistic about the few with no conscience.
- Encourage people to work above their present level.
  - Mentor and watch how they do.

- Reward those who are successful without waiting to be asked.
- Understand that different people are motivated by different means, but the most common motivator seems to be a sense of purpose and that one's actions have impact and meaning.
■ Reward failure in the pursuit of opportunities.
  - Instill the idea that if no one took informed and calculated risks there would be no progress in anything.
  - Demonstrate that the most worthwhile gains result from persistent effort and adaptation to evolving situations because one learns more through trial and observation.
■ Openly acknowledge the importance of emotion in decision making.
  - Help bring people's demonstrations of their emotions more into the open as a valid means of expression that helps clarify motivations and possible constraints in achieving objectives.
  - Know that sharing emotional feelings can help generate and foster stronger trust.
■ Encourage multimedia, immersive virtual environmental and social network interactions.
  - Provide the technical facilities and moral support to enable them.
  - Stay out of the way; do not snoop, but wait for people to surface ideas and solutions that gel through the creativity of the group.
■ Become well versed in mixed and dual reality states.
  - Gain insight into the associated psychological behavior as it may affect people's mental health, views of each other, and their creativity and productivity.
■ Try to increase cumulative trust by integrating or expanding on direct trust-building episodes during periods of limited interpersonal contact.

Chapter 4

Collective Group Dynamics: A New View of High-Performance Teams

B. E. White

Introduction

"Collective Group" in the chapter title denotes an aggregation of people that share some common interests and have energy around delving into a given set of topics. Members of such a group may be loosely or tightly connected. They may or may not have consciously volunteered to be part of the group. They could be part of an outside-directed team focused on a well-defined objective, but not necessarily. Actually, we are more interested in exploring self-organizing and self-directed groups that pursue emergent (not imposed or predefined) objectives, and highlighting their collective behaviors.

Our thesis is that such groups can emulate and even surpass the achievements of what are normally considered high-performance teams. Indeed, this resonates with the findings of Page (2007) related to cognitive diversity. If enough people with different perspectives and distinct ways of thinking collaborate, great things can happen. Such groups often outperform teams of experts, for example, in being more innovative or creating better solutions. This is called *collective intelligence,* a phenomenon that has little to do with the intelligence of the individual members of the group (Johnson 2010; Malone 2010; Wolley et al. 2010).

Much is known and has been written about teams of knowledge workers,* the main class of people that is the focus of this chapter.† We first review some characteristics of traditional teams, primarily for exhibiting a frame of reference for later excursions. Then by way of contrast we explore some of what can be different in groups operating in decentralized (or distributed)‡ environments, particularly those where there is more individual freedom in pursuing creative ways of performing work. Then, we envision a group environment dominated by virtual interactions. Finally, we summarize the resulting implications for leaders of decentralized organizations, especially those embracing organizational learning, as they envision planning for high-performance virtual teams of the future.

---

* A *knowledge worker* is anyone who is employed in the gathering, processing, creation, and dissemination of ideas, information, and data, using whatever relevant media, devices, and tools are available to them.
† We are interested primarily in the group dynamics exhibited by such teams and how their team performance can be improved. The characteristics and principles of high performance in other types of teams are also relevant, of course.
‡ In this book *decentralized* is intended to connote distributed but with a management flavor, i.e., what the organization considers to be a physically noncolocated array of knowledge worker subgroups, some of which may consist of a single member. Thus, *distributed* is used in the usual, more general way, to mean decentralized but with*out* the management flavor. Decentralized organizations may still possess some centralized functions such as corporate or organic financial control and a hierarchical reporting structure. However, such organizations often see advantages in forming distributed teams, going to the best places to find the expertise that is needed.

In this chapter our intent is to provide some general background in group dynamics and build from there in the context of complexity theory, complex adaptive systems, and complex systems engineering (refer to Chapter 1) for chaordic* organizations. Because so much of the background material is well known, and has been experienced firsthand by many readers, we have not felt the necessity of providing a thorough list of references for all statements offered.

Characteristics of Traditional Teams

Traditional teams formed within organizations typically have seven key characteristics (Castka et al. 2001):

1. External governance
2. Face-to-face meetings

---

* "By Chaord, I mean any self-organizing, adaptive, non-linear, complex system, whether physical, biological, or social, the behaviour of which exhibits characteristics of both order and chaos or, loosely translated to business terminology, cooperation and competition." (Hock 1995, p.1) Dee C. Hock so wrote in *World Business Academy Perspectives* 9, no. 1 (1995). His explanatory remarks: "It is almost impossible these days to read a business article or participate in a seminar without stumbling over such popularities as 'learning organizations,' 'empowerment,' or 'reengineering.' It is equally common to encounter in the scientific community the study of complex adaptive systems, commonly referred to as 'complexity.' I find it cumbersome to either think or write about fundamental principles underlying both physical systems and human institutions in the terms unique to either business or science. So after grubbing in various lexicons for a suitable word to describe the kind of organization discussed here, it seemed simpler to construct one. Since the knowledge pursued is believed by scientists to lie on the knife's edge between chaos and order, the first syllable of each was borrowed and Cha–ord (kay–ord) emerged." Willis Harmon provided Hock's foregoing remarks in "*World Business Academy Perspectives* Editor's note: This article ["The Chaordic Organization: Out of Control and into Order"] describes a new organizational form that carries within it the seeds of a new organizational culture—a culture that might well spell the difference between a smooth, orderly transition to a salubrious and sustainable global society, and the chaos and anarchy that some see in our near-term future. I believe this is one of the most important articles we have published to date."

3. Stated purpose on a specific topic
4. Planned accomplishments within a given schedule
5. Funding support but insufficient resources
6. Constrained discussions
7. Insufficient levels of commitment

Item 7 above is particularly telling in explaining only partial team successes. Typically, organizational teams are formed to serve a corporate need, and they are set up and formally sanctioned a term of reference by management in a top-down directed fashion. The *terms of reference*, whether explicitly documented or informally specified, provide the objective of the team, roles and responsibilities of team members, expected product(s) of the team, schedule, and resources. To help ensure ultimate success in achieving objectives, the team should be involved in discussing, clarifying, and finalizing the terms of reference.

The formation of teams usually is motivated to serve an important management purpose that arose and has become clarified in discussions at middle to upper management levels of an organization. Overarching strategic issues are often generated by the corporate officers or the senior leadership. The typical initial response to such a need is to appoint a committee to study the problem and generate recommendations for solution.* Ideally, at least one representative of upper management participates in some of the committee's key meetings to provide appropriate motivation, guidance, and feedback.

---

* However, in some organizations, executives broadcast an e-mail or data call asking the workforce to weigh in on a particular issue and/or propose ways forward. The approach depends on how the top leaders of the organization view themselves and their role in the organization. There is a lot of scholarly work in leadership around this topic.

Although informal teams or groups* may exist within an organization, they tend to form and disperse in an ad hoc fashion, often without being recognized, appreciated, or leveraged by management. The danger of this informality is that an organization can miss untapped potential that could lead to opportunities for strengthening future vitality and viability, things that are essential for long-term survival in an uncertain world.† To protect against this, staff involved in ad hoc efforts should at least keep their management aware of the existence, activities, and results of these informal networks.‡ Almost as important, perhaps, is sharing knowledge, without attribution of course, about the team's behavior from a group dynamics point of view. This can facilitate the spread of good group dynamics practices§ as well as elicit help to correct dysfunctional behavior. All this is described further when discussing distributed team models later in this chapter.

---

* Groups and teams have different definitions in the scholarly literature. There, essentially, teams are formed intentionally to achieve a specific goal (produce a product, win a game or proposal, etc.), whereas groups are simply aggregations of people. This chapter (and book, for that matter) is not intended to be a scholarly work, but more of a useful guide for the practice of systems engineering and program management. As already stated, in this book we view groups as more than an arbitrary aggregation of people, but rather a more informal collection of knowledge workers who interact (at least) loosely with each other and their environment(s) around some common subject, topic, concern, or issue. In this sense, our groups tend to be self-organizing and self-directed.

† As claimed earlier, some informal groups are more powerful and effective organizationally than their formal team counterparts. Some of that is because group members often are self-selected and self-motivated. Also, the existence of the group may not depend on a particular champion or sponsor. In contrast, teams sometimes are short-lived because the champion/sponsor leaves his/her position, and no other leader/manager wants to perpetuate the previous agenda, in favor of doing something different for which they can take credit on their watch.

‡ However, scholars in the field of organizational theory, for example, are very concerned about the ethics of studying and disclosing the existence of informal networks. There are important considerations of human-subjects research and organizational effectiveness at stake.

§ The importance of understanding human dynamics (in relationship to more effective prosecution of our wars) was recently emphasized even by the US Department of Defense (DoD-AT&L [Acquisition, Technology, and Logistics] 2009).

Enneagram

Usually the traditional (command and control type) team constrains their discussion by focusing on only three aspects of what is relevant, i.e., the organizational structure/context, the issues, and the work. Six other aspects that need to be addressed for ultimate success—identity, relationships, information (foundational self-organization domains) intention, principles and standards (quality), and learning, as suggested in the highly effective group discussion process called the *enneagram*—are largely ignored, at least explicitly, not only by the team but also by their governing management as well.

See Figure 4.1 and the explanation of the enneagram, a proven process for productive group discussion in Chapter 5. In Figure 4.1a, the recommended order of discussion of topics interconnected by the black and gray arrows is Topic 1 (Intention), Topic 4 (Principles and Standards), Topic 2 (Issues), Topic 8 (Structure/Context), Topic 5 (The Work), Topic 7 (Learning), and repeating as necessary cycling to Topic 1, etc. This order corresponds to the digital expansion of the fraction 1/7th. Identity has the label 9 as well as 0 to suggest a possible continual upward spiral in the levels of discussion. The

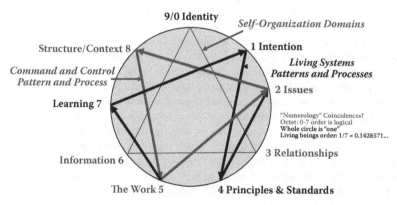

Figure 4.1a   The enneagram web: a proven process for productive group discussions. Adapted from Knowles 2002. Used with permission from the MITRE Corporation Copyright© 2011.

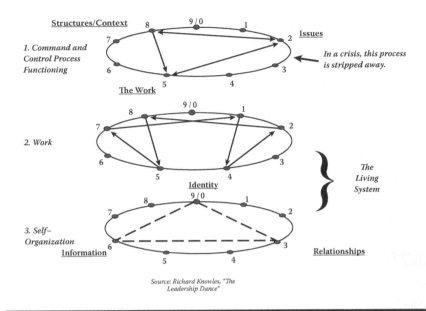

Source: Richard Knowles, "The
Leadership Dance"

Figure 4.1b   The enneagram layer cake: a proven process for productive group discussions. (Knowles 2002, 38)

dynamics, understanding, and ultimate success of the team's activities should be enriched by fully embracing all aspects of the topic. In Figure 4.1b, the three basic processes that can occur simultaneously are shown as layers. Being somewhat counterproductive in a crisis, the upper command/control layer is deemphasized.

Cynefin Framework

Another well-known leadership framework that can help organizational progress and decision making in chaordic settings is the Cynefin framework (Snowden 2000). Refer to (Snowden and Boone 2007) for an excellent treatment on this topic.

Examples and Level of Commitment

Examples of traditional teams abound. In the Department of Defense (DoD) some are known as Integrated Product Teams

(IPTs).* This term captures two ideas: namely, that some product is expected, and that it is to be developed considering all relevant aspects and in an integrated, holistic way. This IPT concept applies more broadly to other government domains (Creekmore 2008) and to commercial industry as well, and may be known by other names. In all teams, at least an acceptable level of healthy group dynamics is necessary for success (Wikipedia 2011a).

Ideally, members of IPTs are expected to be able to act autonomously and with the authority of the organizations. To the extent they do so, the team performs better, more effectively, and more efficiently. This is primarily because much time and backtracking can be saved by not requiring continual coordination with all the respective organizations between meetings, especially those not associated with the critical path of work flow.

Team members† are usually assigned by management based upon their technical expertise, agreeableness (e.g., the extent they are known to be a team player), and availability.‡ The latter factor can be critical in that most managers typically want to retain their best people for work on the perceived relationship of their suborganization's own most pressing problems. Thus, the team may contain members that are not ideally suited to the purpose, unless the team purpose maps closely to the represented organization's objectives. Sometimes members are offered as pro forma representatives, attending meetings primarily to inform their bosses as to what's happening and to help keep their parent organization out of trouble or to minimize its risk. Thus, the best IPT members, from the

---

* Nowadays many IPTs are less traditional in that they often meet in virtual environments.
† Generally, one can ask, "Why does someone participate in a team?" They may volunteer, be assigned, be there to contribute, to protect their organization's equities, to be an informant to their own manager, or to even be a saboteur.
‡ Assignments normally include representatives from peer suborganizations, a characteristic that is often implicit and satisfied as a result of management making sure the appropriate technical expertise is represented.

parochial organization manager's point of view, tend to be the ones who can most effectively comply, or at least appear to comply (creative noncompliance), with IPT actions without jeopardizing the manager's own agenda. Of course from the team's perspective it is better if candidates are encouraged by their bosses to proactively engage collaboratively and creatively pursue solutions.

Again, from an overall management perspective and for team productivity, it is much better if members are made available primarily based upon the importance of the IPT's output, as recognized by all the contributing managers.

Behaviors

Teams usually are not self-directed. There is an appointed leader who is accountable to management for the team's accomplishments. Through this leader, the overseeing manager, for whom the team is to create a product, often tries to exert control over the team, deciding whether to accept results and how to create incentives and allocate rewards (or punishments, e.g., admonishments).*

Traditional teams most often operated on a face-to-face basis, either through colocation or travel. Approximations to face-to-face encounters, i.e., video/audio teleconferences, intended to be good substitutes, have been attempted valiantly for years. More recently, meetings in other virtual environments, notably within social networks such as Second Life, are coming into use, albeit on more of an experimental basis. In any event, it is always a challenge to build mutual trust (refer to Chapter 3) within teams rapidly, even with face-to-face meetings, no matter what the nature of the meeting venue. Face-to-face meetings can be

---

* There is a huge amount of scholarship, theory, and practice around this topic, e.g., much of the popular literature discusses Theory X vs. Theory Y. The most popular ways of thinking about this boils down to carrots and sticks.

especially productive when people have opportunities to tell personal stories to each other (Kahan 2006), during either the business portions of the meetings or meeting dead times. However, as already noted, virtual organizations are becoming more prevalent. Handy (1995) provides a working model for applying six facets of trust to the virtual work environment to ensure bonding occurs among those dependent upon each other.

Effective teams often operate and evolve in accordance with the classic forming, norming, storming, and performing process.[*][†] To generate ideas, brainstorming rules are often proposed and sometimes followed. Some of the ground rules successfully employed in teams include the following:

- Attend meetings or send a representative in unavoidable circumstances.
- Be prompt and mindful of the agenda so as not to hog air time.
- Treat each other with respect.
- Keep discussion specifics within meeting walls.
- If/when the group reaches consensus, don't continue to argue.
- Do not revisit old ground unless there's a compelling reason.[‡]

---

[*] Wikipedia, http://en.wikipedia.org/wiki/Forming-storming-norming-performing

[†] This is a popular way to think about what happens in a team, and people who have been trained to think in these terms usually enact this process. However, the empirical studies show that teams don't go through this process as a series of steps; rather, these are activities that happen iteratively and recursively. Some groups dwell in one area so long they get stuck. Other empirical work has shown that you can get teams working faster and better if you take away this mental model.

[‡] A significant change in the environment surrounding the problem the team is trying to solve may suggest a strategic retreat with an accompanying revisitation of the team's reason for being, stated (or unstated) assumptions, and ground rules.

■ Listen to scuttlebutt remarks as people are leaving the room or speaking in the hallways afterwards as this is often a good indication of the meeting's effectiveness.

High-performance teams focus on the objectives and issues, not personalities, and air many suggestions for process improvement during meetings. They tend to stay focused and not get distracted with gossiping or cliquish grousing during or between meetings. An example of published characteristics of high-performance teams includes the following (Traut 2008):

■ Participative Leadership
■ Purpose and Vision [Alignment]
■ Task Focused
■ Shared Responsibilit[ies]
■ Innovative[ness]
■ Problem Solving
■ Communicative[ness]
■ Responsive[ness]

From a management point of view, the team is golden, i.e., ultimately the team is expected to present unified positions, and team consensus (or with well-developed and understood areas that lack consensus) on every issue. This is implicitly—if not explicitly—understood by team members, and individual behaviors within the team are shaped accordingly. A high-performance team will track areas where the team has differences of opinion, and develop effective approaches to resolving these discrepancies in a timely way. Although not all disagreements can be resolved within the time frame of the team objectives, they are always noted, tracked, and earmarked for further attention.

Group cohesion often results if group members share the same viewpoint, either at the outset, or after a process of discussion, debate, and consensus building.* It is one of the ways group identity is formed and maintained. Dissenting viewpoints threaten group identity and cohesion, and thus are resisted, often forcibly.† The human tendency to down-play ideas not invented here creates dissonance, especially with respect to outliers who may have ideas that don't con-form to group expectations and who may have less influence in the group.‡ One characteristic of teams that do not per-form well is the suppression of dissenting members' views. Sometimes dissenters are attacked personally and are pres-sured to cease and desist, or to depart the group altogether. When this happens, the team loses intellectual content that could have been quite valuable in providing solutions to the problem that led to team formation. Clearly, team members should be sensitive to these situations and work toward pre-vention and cure to the extent feasible within the group, or elevate any remaining issues to management. (There is more on members who have divergent views from the group later in the chapter.)

---

* Many teams experience some degree of conflict, but when the best teams have disagreements, they address them meaningfully and respectfully, and come to agreement on most of the issues. On any particular issue, they may agree to disagree, they may reframe the issue to finesse the disagreement, or they may persuade most people into something approaching full consensus.

† This does not often happen in truly effective teams. Dorothy Leonard and others address the concept of "creative abrasion" as essential for a healthy team (Leonard and Swap 1999; Garvin, Edmondson, and Gino 2008; Handy 1995).

‡ Groups take their cues from their leaders—either the formal leader(s) or the ones who have respected expertise or organizational or social standing. If the leaders enable respectful disagreement, the group is more likely to succeed. If the leaders are incompetent, conflict-averse, or insecure, the group is likely to disintegrate. It will no longer fit the definition of a team and may lose members to the point it is no longer even a viable group.

Dysfunctional teams can be embroiled in political infighting* imposed by the will of the external organizations of the team representatives. This happened in the mid-1990s, for example, in an RTCA committee wrestling with establishing a standard for a new air–ground communication system for air traffic management. At the outset, powerful organizations that were aligned with the interests of the airlines were pushing a particular waveform implementation, while those organizations aligned more with the interests of the Federal Aviation Administration (FAA) were proposing a class of alternative waveforms.† Fortunately, over time and after much respectful debate during many face-to-face meetings,‡ the RTCA team reached consensus on a recommendation for adoption, and ultimately an International Civil Aviation Organization (ICAO) standard was established (ICAO 1996).

Team building events are geared toward creating amiable relationships and trust. It is significant that these activities often are separate from the business-oriented team meetings. But in the government acquisitions world, for example, some aspect of the business must be cultivated during such events to justify to government bean counters the monetary expense

---

* There are many ways a team can be dysfunctional. For example, the formal authority (i.e., the appointed lead) may not have the skills or social standing to actually lead. Others in the group may contend for positional leadership or recognition. Ordinarily affable and compliant people may see their role symbolically as representing something bigger than themselves; their behavior may be disproportionate or inappropriate to the task. Teams may not have and adequate variety of skills and competencies, or they may be so homogeneous that they agree but their recommended solution is unacceptable to other unrepresented stakeholders.

† Much of this work is documented in the records of RTCA Special Committee (SC) 172. VHF Air-Ground Communications System Improvements Alternatives Study and Selection of Proposals for Future Action.

‡ Time pressure to increase the communications capacity of bandwidth-limited channels provided significant impetus. This, despite the fact that the impending doom of running out of channel capacity still looms many years later! By 2011, the emphasis switched to NextGen air traffic modernization, driven by a Global Navigation Satellite System, e.g., Automatic Dependent Surveillance – Broadcast (ADS-B), http://www.insidegnss.com/node/2582

associated with what they may view as simply having fun.*
High-performing teams build relationships as they engage
with each other, doing the business of the team, and/or during
storytelling, while generally enjoying the encounters. In such
cases, separate team-building events may no longer be necessary or productive. They can be used instead as opportunities
to celebrate team success.

Distributed Team Models

Increasingly, the team is geographically dispersed and distributed throughout the information space, and is reliant on multimedia modes of communication, such as e-mail, shared web
spaces, video or audio teleconferencing, immersive 3D environments or virtual worlds, and even social networks such as
Twitter, as well as occasional personal visits.† Team members
continually join and depart the overall group.‡ Because of this

---

* The issue of appropriate use of funds is a fundamental issue of governance and
stewardship in *any* public or private organization, and bean counters are powerful in many places. Some nonprofit organizations, such as Federally Funded
Research and Development Centers (FFRDCs), may be more scrutinized than
most. Instances of abuse that become public, like AIG's management who went
ahead with their company retreat at a posh spa a few days following their federal bailout in October 2008, can be splashed all over the news media. This type
of thing brings even more negative publicity to a troubled organization.

† A geographically distributed team can be just as focused in purpose and activities as a geographically homogeneous team. But to some degree they tend to be
hampered by fewer opportunities for face-to-face communication and lowered
communication bandwidths.

‡ In hierarchical organizations, in particular, where top-down-driven teams are
more prevalent, and managers want some degree of predictability of team
behavior and team outcomes, this characteristic tends to be an undesirable but
still largely unavoidable trait of teams. External events often dictate individual
departures from teams and the need for replacements. In such situations this is
a governance issue that must be managed continually to ensure team cohesion.
On the other hand, in more decentralized and distributed team models, team
membership is more dependent upon and governed by individual decisions
surrounding issues of personal motivation, interest, comradeship, and feelings of
contribution.

dispersion and the greater likelihood of ad hoc participation, the overall purpose can be fuzzier. We hypothesize that group objectives and results may emerge more through the interactions of team members in a self-organized fashion than from specific requests driven by management. Accordingly, the creation of desired outcome spaces* is arguably more difficult in distributed environments. Something that might help is for managers to more strongly empower their distributed knowledge management workforce.

Enlightened managers create appropriate incentives† for their teams and team members and conditions that enable team self-organization. They adopt the philosophy that the answers are in the room (Oakley and Krug 2006). To a large degree, the team can be self-organizing, although it is unlikely the team will determine its own rewards and punishments.‡

One thing to ponder in carrying this distributed model to the extreme: What if team membership is completely open

---

* We use *desired outcome space* as a name for the description of an abstract, multidimensional realm covering the entire set of possible outcomes sought as result of a team effort. We claim that time spent elaborating on outcome spaces, rather than diving too early into the pursuit of specific outcomes, will pay dividends in greater achievements satisfying overall objectives. This is akin to realizing a good return on investment (RoI) in the application of the systems engineering discipline in a development program (Boehm, Valerdi, and Honour 2008).

† Incentives that reward true results, i.e., those that contribute potential solutions in the desired outcome space, are more appropriate than those that protect or enhance the relative importance or well-being of participating individuals or organizations, for example.

‡ Again, top-down-driven teams inspired by a control mindset are not self-organized; they are formed quite intentionally by management directive. In such cases, the team tends to be governed from the top to ensure that the expectations placed on the team when it was formed are met. Here the rewards and punishments of the team come from external sources, e.g., the stakeholders expecting results.

based upon individual self-interests?* Would this enable the full potential of self-organized solutions characteristic of adaptive, robust complex systems of independent actors or agents?† We believe that this extreme could ultimately prove to be the most fruitful in creating desired solutions. However, at least a modicum of governance, provided by some portion of an organizational hierarchy, is usually beneficial for maintaining focus for the team. (Refer to Chapter 1.)

The reality though (as experienced firsthand by one of the authors [White]) is that most staff members supporting organizational goals do not have complete freedom to work on whatever they want.‡ Some are even constrained so much that they must justify every hour they work to their management to assure them that they are only working on preapproved projects or tasks.§ One would hope that most managers use common sense, however, and allow staff flexibility to loosely interpret their direction to include the exploration of subject matter that can benefit their main tasks, either indirectly now or directly in the future.

---

* Many traditional thinkers would say, "Then you do not have a team!" They may feel strongly that a team has to be channeled and focused (by an external authority). Here we are taking a broader view of the definition of team that admits collective individual motivations as the basis for team cohesion and the inherently interdependent nature of interactions necessary to achieve the team's goals. Of course, externally governed teams can also be creative and engender free-flowing thought.

† Indeed, self-organization is a characteristic of volunteers participating in list servers (listservs) and online communities of interest (CoIs). They can be very enthusiastic in addressing specific issues and problems, and sometimes shedding light on potential solutions and potential ways forward. Thus, in this sense they can serve as a team now and then.

‡ There is nothing harsh about this reality! It is business! It is life. The management challenges are to accept, foster, and support creative environments in the workplace that include the likes of Google, Wikipedia, virtual networks, e.g., Second Life, and social networks, e.g., Twitter and even Facebook!

§ For example, some Contracting Officer's Technical Representatives (COTRs) are known for considering posting to a technical issues board to be time off-task, even though such actions reach out to others who might have already solved the problems faced on the subject programs.

Serious Games/Virtual Worlds for Training

It is important to consider the many ways for improving how distributed teams get to know and trust (refer to Chapter 3) each other. One can try to bring most team members together in a face-to-face formative off site location dedicated to the purpose. If this is deemed impractical for reasons of travel cost, for example, one could try a video teleconference (VTC). JumpStart Storytelling (Kahan 2006) can be tried in a virtual environment. However, there are other means of training distributed teams in immersive 3D simulated environments so they can learn how better to interact with each other as if they were face to face.* An important concept to consider harnessing in distributed environments is the idea of utilizing serious game† and immersive 3D environment technology such as Second Life (2008) and Google's newer 3D virtual experience website service called Lively (Google 2008 which was discontinued on 31 December 2008, http://en.wikipedia.org/wiki/Google_Lively).‡ Serious games have progressed to the point where individuals can interact in a virtual environment but with many of the characteristics of face-to-face encounters that take into account speaking and hearing, facial expressions, and other body language indicators. Each individual is represented by an avatar (usually, but not necessarily, a human-like

---

* This notion is trying to achieve much more than just putting ping pong tables in the workplace, etc., but admittedly leaves open to future applied research just how much this would affect team behavior and cohesion.

† In this age of the Internet, one may tend to associate games with online video games focused on violence, evil pursuits, sensationalism, and other (at least perceived) sinister purposes that seem to attract children or young people in our current US culture (Webster 2008). As described in this chapter, serious games are referenced as being used for collaborative and learning experiences.

‡ One might go even further and utilize such media even in face-to-face meetings! With the prevalence of high-frequency texting of personal information amongst teenagers using iPhones and iPods, for instance, some teachers are creatively leveraging and integrating this addiction into the conduct of classroom instruction.

image)* that can be created to suit the subject's desires as to the type of image he or she wants to project. Movements of the avatar can be preprogrammed adaptively and/or controlled as prescribed by its owner. Individuals interact in near-real-time using animation scripts of varying degrees of prespecification but in an adaptive way, depending on what the other avatar is saying and doing. It is also possible to interact with an unknown group of avatars in open public spaces not behind enterprise firewalls. In these cases there are general rules of etiquette established by the virtual world culture that governs interactive behavior.

Since at least 2008, the US Air Force has experimented with and adopted serious game ideas in its recruiting and training program, for enlisted personnel in particular (Hughes 2008). Figure 4.2 shows a wealth of existing multimedia capabilities available to today's knowledge worker.[†] Who would be much surprised if this trend continues well into the future?

The Air Force adopted implementation goals using MyBase *concepts* such as in the following list.

- Embodies a virtual, exploratory, and interactive *environment*
- Supports a *mix of live, virtual and constructive* education and training
- Provides *learner interface* access to AF knowledge bases and management systems

---

[*] Significant scholarly and practitioner work exists on how closely an avatar should resemble its protagonist. (Yee, Ellis, and Duchenaut 2009; Yee et al. 2007; Yee and Bailenson 2007; Castronova 2003). Refer to the first citation for a creative mindset that emphasizes the importance of considering a whole new [virtual] world available to us if we do not insist on re-creating approximate physical replicas of ourselves and real objects in virtual worlds like Second Life.

[†] Of course, the full array of these type capabilities is rarely available at the desktop of every knowledge worker, or even necessary for doing the job at hand! However, perhaps with such an array of possibilities, with more to come, the job description of future knowledge workers may change significantly. This type of environment could well become the office of the future and make working from home even more prevalent than it has become in the twenty-first century.

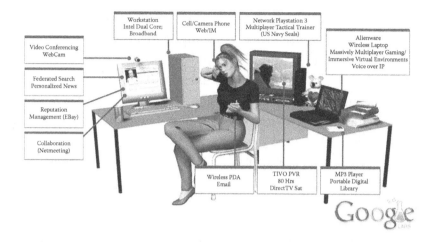

Figure 4.2 Today's cyber teenager and tomorrow's knowledge worker (From Hughes, Larry. 2008. "On Learning: The Future of Air Force Education and Training," Chart 6. Presentation at Organizational Learning Technical Exchange Meeting II. The MITRE Corporation. April 22. This information is in the public domain and is available at the following URL: http://www.aetc.af.mil/shared/media/document/ AFD-081216-008.pdf).

- Supports *continuous*, lifelong learning
- *Integrates* existing systems
- Enables *precision learning*
- Includes platform for *collaboration*.

General William R. Looney, III, former commander of the Air Education and Training Command (retired from the US Air Force on 1 August 2008, http://www.af.mil/information/bios/ bio.asp?bioID=6229). Hughes (2008) characterized this Virtual World concept* as follows.

---

* In 2008 MyBase and Virtual World were concepts yet to be proven, although the Air Force was clearly trying to establish them in their training environments. Details regarding the challenges in the implementation and benefits they have achieved remain to be seen and are worth tracking. In May 2011 the DoD announced it was using virtual world simulations such as Enhanced Dynamic Geosocial Environment (EDGE) to help train today's warfighter for real combat situations. Refer to http://www.vr-news.com/2011/05/13/ dod-explores-virtual-worlds-for-military-training/

1. Shared Space: ... many users [can] participate at once.
2. Graphical User Interface: ... ranging ... from 2D cartoon imagery to more immersive 3D environments.
3. Immediacy: ... world allows users to alter, develop, build, or submit customized content.
4. Persistence: ... existence continues ... whether [or not] individual users are logged in.
5. Socialization/Community: ... allows and encourages ... formation of ... social groups.

Thus, we believe individuals can be trained in virtual environments to operate effectively within teams.* Various scenarios can be simulated to reflect real situations that are likely to appear in team environments associated with the classical stages of forming, norming, storming, and performing. Of particular interest are processes for generating interpersonal trust and the handling of individual dissenting behavior (refer to the next subsection). This could be developed as online training with various modules that team members could exercise, especially upon joining the team but also when needed to refresh needed awareness of psychological factors that often drive team behaviors. Facilitators and trainers could update the modules based upon important incidents experienced by the team. It would be interesting to record individual feedback.† Better yet, this might be done with well-designed interactive software including video recording of facial expressions, etc., while taking different aspects of the training to assess the subject's relative sense of comfort or well-being. (Theory suggests that the brain generates oxytocin in greater quantities when one is feeling the trust emotion (Health E-Learning 2005 [refer to Chapter 3]).

---

* In this chapter we are exploring what training techniques are needed to do that; clearly, we do not yet have firm recommendations on necessary ingredients, only a preferred way forward.
† It would be best to conduct a formal assessment of the training events to measure improvement in effectiveness.

Direct measurement of that, however, is probably impractical; oxytocin may be correlated with feeling good, but would not be a *predictor* of trust.)

Difficulties in Distributed Environments

In distributed environments, isolation can be a problem, and we believe in creating conditions for minimizing isolationist tendencies, if we can, particularly in virtual team settings (Chown 2007). It is pretty well accepted that the old Monroe Doctrine, an isolationist policy for the United States, was a mistake.* Similarly, there is that Machiavellian saying, keep your friends close but your enemies closer. The latter policy is probably more beneficial in the long run. Isolation of parts of the world, a country, an organization or team, or of individuals, can create misunderstandings and resentment among those isolated that eventually can lead to serious problems for the whole. Thus, isolation is best avoided.†

Achieving cohesiveness of the team, if that is an important goal, is more challenging in a distributed environment. But there are many documented strategies for dealing with this, including setting up conditions for interconnection and frequent interchanges or interactions. Thus, a team is apt to be less cohesive in a distributed environment, but that is not

---

* At least this is what was taught in one reputable Midwestern high school in the fifties. However, many might agree that our US culture has a strong isolationist streak even now; we seem to veer between wanting to be left alone, and getting into overseas entanglements.

† This paragraph may stimulate some philosophical arguments. There are many reasons to encourage interaction, and to guard against excessive isolation. However, an individual (like the famed movie actress Greta Garbo) may assert a right to be left alone. The amount of contact a person needs to have to feel included and important varies depending upon the individual and the nature of the project or task. Some people need a lot of contact, others like incidental contact to just keep the channels open, and still others only need occasional contact.

the end of the story.* Maybe *team* is even the wrong word, at least until the participants interact sufficiently well to engender trust among them. In such cases there are probably even fewer attempts to organize neatly. Compared to traditional teams, in a distributed environment subgroups form more dynamically and work more semi-independently and in parallel, and in a more ad hoc fashion.† On the other hand, distributed teams can be quite powerful and undeniable by an opposing force that is organized hierarchically (Brafman and Beckstrom 2006).

In distributed environments where activities may tend to be less cohesive and more uncoordinated, there would seem to be a greater likelihood for disruptive event generation or emergence. It is essential to identify and resolve such disruptive incidents because they can be highly significant for either good (e.g., as creative force for spreading innovation) or ill (e.g., as something to suppress or mitigate in its propagation). It would be interesting to investigate whether disruptions‡ in distributed networks of knowledge workers follows a power law (Wikipedia 2008b). Namely, what if the probability of a disruption along with the impact of that

---

* Kerry Buckley's research (Buckley et al. 2009) tended to show that groups that establish face-to-face relationships and trust will use whatever technology they need to stay in touch. Some research into communities of practice (CoPs) and CoIs shows that people seek out online groups so they can have a sense of belonging, or an opportunity to develop themselves through learning and growth. But these groups, even if they are communities, may not be *teams* in the most common scholarly use of the term.

† Interdependency among distributed groups can be a characteristic that management of a control mindset feels is essential to manage. In this case, ad hoc-ness would be a sign of poor management of the distributed team, not a desirable characteristic. On the other hand, encouraging ad hoc interactions can stimulate creativity through diversity and lead to innovations beneficial to the organization.

‡ There are many documented ways a group can be disrupted. Some are the same whether the group is face to face or distributed. For example, strong political opinions that are not germane to the group's purpose, egocentric or un-self-aware behavior, violating group norms, etc., can be disruptive regardless. In addition, there is relevant scholarship on behavior within groups, e.g., the MITRE Corporation and Johns Hopkins University have conducted some research on phase-change behavior (Booker and Strong 2008).

disruption, e.g., degree of spread throughout the network of a controversial idea that exercises others, follows a power law? This phenomenon has been observed in warfare, especially that of the terrorist ilk (Highfield 2008). Of course, disruptions can have positive as well as negative effects. Disruptive technology, for example, though potentially having a devastating effect on the future of related existing technologies, can ultimately be quite beneficial to everyone as an incoming tide that raises all boats.

Creating conditions for open discussion in distributed environments is an old and very rich topic. We cannot think of a better way to facilitate this goal than with the introduction of the enneagram process (refer to "The Enneagram" under "Characteristics of Traditional Teams" earlier in this chapter). In a healthy environment, which admittedly is perhaps more difficult to achieve in a distributed organization, minority views are encouraged and respected. The *ideas* are subject to debate, but the people who propose them are not attacked, and their ideas are heard. Everyone needs to be conscious that we all see things a little (sometimes a lot) differently, and that no one has a complete understanding of the true underlying reality (McCarter and White 2009). As O'Connell (2008) powerfully points out, each of our views of reality is based upon our beliefs, which are in turn derived from myths (common mental representations of our world) that have been generated, passed on, and evolved over long periods of time. Consequently our *perceptions* of reality, and not reality itself (an abstract concept), govern our actions in whatever we do including working in teams, whether they are closely coupled such as IPTs and some project teams, or loosely coupled distributed teams. As O'Connell suggests, an idea should not be viewed as true or false but rather as something that should be challenged, i.e., examined and tested objectively.

As in any kind of knowledge-worker team, conflicts over ideas are to be expected. Some best practices for trying to handle such conflict* are espoused in a mentoring guide available on the Internet (Triple Creek Associates 2004).† Although these tenets are written from the point of view of the person being mentored, they apply more generally to how one should behave in teams.

Withhold Judgments
[K]eep an open mind ... Use I statements ...

Speak Precisely
Be specific ... reference present circumstances ...

Remain Considerate
Maintain a polite frame of mind and attitude. ...

Focus Your Discussion
Clarify points of agreement before dealing with disagreements. ...

Balance Your Communication
Express your thoughts and needs ... in a balanced manner. ...

In any team, either the more traditional or distributed variety, it helps productivity if the team and its overseers understand that one only learns from mistakes‡ (Ackoff 2004). One needs to include failure as not only expected but invited behavior (Leonard and Swap 1999; Handy 1995). Google also is an example of a flourishing business that bases their environment on failure (Hammonds 2007). If team members are overly cautious and too critical of proposals for action—for

---

* Conflict management is essential, but this isn't just a skill for the formal team lead to know. Conflict is often best managed by peer interaction that can keep the instigator engaged in the group so their knowledge and distinctive contributions aren't lost. See the later subsection "Exploration of Dissenting Individual Mindsets."

† There are many other practices available in the academic literature, e.g., Leonard and Swap 1999; Handy 1995; Garvin et al. 2008.

‡ This means that when you are performing without making mistakes, you already know what to do in the present context, and therefore you are not learning. Of course, we also can learn from reading, observing others, watching what works, hearing about *their* mistakes, etc.

example, because of a tendency to risk aversion—very little will get accomplished. This is consistent with the Forward Focus of Oakley and Krug (2006). As they point out, fear is the ultimate *backward* focus. (For more on fear, refer to Chapter 3.) Informed risk taking and the pursuit of opportunities are what engenders eventual productivity.

X-Teams*

Ancona and Bresman (2007, 6, 79–80) state that membership in an X-Team "with members working outside their boundaries as well as inside them"† takes "courage and determination. ... Sometimes top management simply does not want to listen to new ideas. ... There is a fine line between going after what one truly believes in ... versus continuing to argue. ..."‡

Also, as Ancona and Bresman (2007, 2, 6–8, 65, 129) point out, many teams with high potential ultimately fail primarily because they are too inwardly focused. Their X-Teams are strongly externally focused. They espouse the following elements/principles in their recommended X-Team framework.

- External Activity
  - Scouting
  - Ambassadorship
  - Task Coordination

---

* The use of X-Teams cited here is only one of several possible approaches for overcoming difficulties in distributed environments.

† If you do not understand what this means, please read on. But this is a standard idea for the team to consider, i.e., how to balance attention to the problem at hand and the external focus.

‡ This paragraph goes to the issue of how well groups and teams fit within the prevailing organizational and leadership cultures. If the organization's formal leaders want rubber stamps on ideas and decisions, then they won't delegate authority, responsibility, and decision latitude in proper proportion for really creative work. That may be okay with the team members, as long as that's what they expect and understand. If the team assumes that it actually has authority and responsibility but finds out that's not so, it is a recipe for mutiny or creative nonparticipation.

- Extreme Execution[*]
  - Psychological Safety[†]
  - Team Reflection
  - Knowing What Others Know
- Flexible Phases
  - Exploration
  - Exploitation
    - Visioning
    - Inventing
- Exportation

The carrying out of these principles is guided by their X-Factors.

- Extensive Ties
- Expandable Tiers
- Exchangeable Membership

The result is an X-Team whose members frequently navigate across the team's boundary.

Regarding Psychological Safety, there is an excellent nearly-hour-long video[‡] with Jack Welch, former head of General Electric, and generally accepted as an outstanding leader, motivator, and corporate chief executive. He is very convincing, describing in great detail with examples, what it takes for success in business: high expectations for pursuing opportunities

---

[*] Participatory and transparent decision-making procedures are encouraged, including "Heuristics—or rules of thumb—that give guidelines about and boundaries around the process and help team members make decisions when circumstances are ambiguous. ... 'no one has to sit through an entire meeting.' ... 'When someone was stuck on a problem, we didn't want them to wait more than two minutes to ask for help'" (Ancona and Bresman 2007, 107, 109, 111).

[†] "'psychological safety' means that all members feel the team is safe for inter-personal risk taking. ... success has ... been attributed to a culture in which ... teams feel free to talk about their own mistakes and those of others without fear of punishment. ... A culture of extreme execution requires learning as you go, ..." (Ancona and Bresman 2007, 92–93, 97).

[‡] MITWorld 2011, "New Conversation with Jack Welch," April 26, 2011, Running Time: 0:57:40.

with vigor and learning from mistakes after taking informed risks. No one is punished for attacking but not achieving a worthwhile goal immediately. This leads to a vigorous and healthy work environment, profitable innovations, high product quality, and employee and company longevity. Those that don't measure up to such standards are told early and often; if they can't cut it, they are encouraged to leave or are dismissed.

Gawande (2007) gives some examples of how change really happens. Being a surgeon, mainly working in hospital environments, he details how difficult it is to establish, but ultimately how greatly satisfying it can be, when one is able to institute a rigorous but demonstrable healthy practice of thoroughly washing one's hands after seeing each patient. This takes persistent, persuasive, practical, and visual reminders to doctors and nurses of what is necessary. Similarly to the practices advocated above, lasting results are easier to achieve by discovering what someone does that is successful, and then helping them amplify that (also, see Oakley and Krug 2006), than by trying to impose some abstract or theoretical technique from the outside that we, as perhaps unwelcome advisors, think is expected to work. Most professionals resent outside help on general principles, feeling that they know best. Thus, first and foremost, one should adopt a healthy degree of humility when confronting complex problems. This promotes openness to new or better ideas from others. Humility can also help smooth the way when dealing with individuals within your own team and within other groups.

Exploration of Dissenting Individual Mindsets

Ideas that surface in a team environment can vary considerably. Indeed, that's what makes a team's outlook potentially stronger than that of any individual. As members of the team interact there is a natural tendency to coalesce many of the ideas into mainstream themes and to discard those ideas that don't seem to fit.* In general, this is fine, provided enough discussion is devoted to weighing the merits of the discarded ideas, along with the proviso that they can always be recalled if further consideration warrants. Such a process is one characteristic of high-performance teams. However, some ideas not yet part of such groupings persist because of the insistence of their proponents in advocating their adoption. These proponents are what we consider to be dissenting individuals.

Some of these individuals are considerate of the feelings of the rest of the group and argue their case with respect and objectivity. But it is not uncommon that some can have the kind of personality that tends to go beyond the merits/demerits of their idea, sometimes questioning the integrity of other group members. This can lead to such an outlier being ostracized from the group either overtly by being shouted down, or covertly by being ignored, for example.

High-performing teams will have a strong code of conduct to prevent, or at least minimize, this type of behavior. In top-down-driven teams it is sometimes up to the leadership to help ensure that what is described here does not happen. This can be done in various ways, including explicit listings of expected desirable and undesirable behavior, and group dynamics training in neutral settings, preferably before the team is formed. In more decentralized and distributed environments, the group itself can, over time, establish desirable norms that will model appropriate behavior and discourage destructive departures.

Another tactic that can be employed by the team leader/facilitator about a dissenting individual's idea is to approach

---

* This depends a lot on the organizational and/or social standing of the person who pitches the nonstandard ideas. If that person is revered as a thought leader or has been an organizational hero, their ideas may be adopted even though they're novel or in the minority. If the proponent has the ear of a big boss who chartered the group, their ideas may get undeserved weight because the group may figure there's no point in trying to put forward something that this connector doesn't espouse.

some trusted member of the team privately and try to sell him/her on the potential merits of idea. If the trusted member agrees, then at the next opportunity, s/he espouses or supports the idea to help achieve greater acceptance by the rest of the team. This may work better than the leader supporting the idea directly.

Unfortunately, group leaders can be guilty of this sort of behavior, as well. At times, at least, some group leaders will shut down others in the group who don't go along with what they want, and what they get the rest of the group to agree to do. Therefore, such group leaders can also be seen as a dissenting individual, i.e., a negative form of outlier.

When all the ideas of the members of a group are not allowed to be heard with respect and courtesy, the result can be dissention among the group, passive aggressive resistance to ideas that some members were forced to accept, group think, where important information is ignored perhaps resulting in negative consequences. Thus, such groups do not work well together.

In a high-performing team, dissenting individuals are made to feel valued, and appropriate social skills for interpersonal interactions are evident. Unfortunately, the latter seems to be on the decline, in general, with the advent of more online interactions. When all team members embrace and practice an enlightened viewpoint, that is to say, that the collective sharing, discussion, and processing of diverse ideas can lead to better overall understanding, it is more likely that minority views will be heard with respect. By further suspending tendencies such as the "not invented here" and "I don't like you" mindsets of some individuals and groups, members with thoughts out of the mainstream truly will be heard.*

---

* To some readers this may sound like we're talking about kids on the Internet. However, some may also recognize that this kind of behavior does occur, unfortunately, in corporate and professional meetings. The ability of external management—or the leader and the group—to create the right team environment needs to be emphasized.

As already stated, some of these dissenting individual outliers may have a lack of social skills for interpersonal interactions. It is important to separate that possibility from the outlying thoughts themselves, and consider these ideas on their own merits or demerits.* This can be an iterative process. Initially, an outlier might try to shock the group with what he or she might feel is a particularly provocative idea that they know will be immediately rejected by the group. Wouldn't they be surprised if the group treats the idea seriously and gives it due consideration, although it still might not survive the discussion? When this happens, the submitter might be encouraged to continue participating with other ideas, perhaps equally creative but maybe not so shocking and presented in a kindlier way. Conversely, if the group is not hip to some of the vagaries of individual behavior, they might attack the submitter at the first instance and discourage their future participation, thereby creating passive aggressive resistance of future group efforts.

It is useful to remember that not all ideas are good ones. Typically many must be offered to be assured that some really useful ones will appear eventually, often after shaping through reasoned group discussions. It's curious how many people are incapable of adhering to even the simple rules of brainstorming;† the natural reaction is to attack an idea that does not fit well within their own belief system. A high-performing team will behave better consistently and treat all

---

* Again, the team should be able to (1) recognize outlying ideas and have a process for dealing with them, and (2) distinguish an outlying idea from a person who likes to be the contrarian.

† This reflects the dominant personality types of typical analytical and engineering people, many of whom are inclined to go deep and have a hard time generating a contribution spontaneously. There's also a generational component; Generation X (born in 1965–1980) and Generation Y (Millennials, people born in 1981–1999) (Lancaster and Stillman 2002, 13) kids were reared with different attitudes regarding competition and cooperation than were older folks, i.e., the younger kids often were taught they were all winners and consequently when disappointments arise, don't know how to accept losing gracefully.

members with respect, actively listening to and encouraging the free flow of ideas. A multiplicity of opinions, thrusts, and activities is not only acceptable but encouraged.

Program/Project Management

Thus far, we have not explicitly mentioned program/project management in the discussion of groups and teams. One can appreciate that teams, de facto or otherwise, are most commonly inherent within the staff compositions of ongoing projects or programs. So what can we say about the relationship of program/project management and teams that has not already been broached? Probably not much other than to (1) emphasize that program managers and project leaders need to consciously pay attention to, understand, try to shape, facilitate, and leverage the group dynamics that inevitably occur among their program/project personnel; and (2) consider how to balance the need for providing specific top-down directions to teams versus the potential benefits of permitting the wisdom of crowds (Surowiecki 2004; Page 2007) to bubble up. We do not intend to delve into the many other issues of this huge topic of program/project management.

Generally, a program has a broader scope than a project, especially if a program includes several projects. But internal to an organization supporting an external program manager, a project can be used as an alternative name for the program. A reasonably sized project usually is composed of several separate tasks as well, but we will not delve into tasks because essentially all of the principles we espouse apply to task management as well as program or project management.

To obtain good results in program or project work, it is appropriate to focus on, among other things, facilitating team communication, cooperation, and collaboration. Many of the principles of eXtreme Project Management (DeCarlo 2004) seem relevant here:

- *Adopt a quantum mindset* ... facilitating the flow of thoughts, emotions, and interactions ... in times of high uncertainty, high change, and high stress.
- Shift ... away from ... program/project artifacts and toward ... program/project dynamics over mechanics.
- Create an *environment that fosters innovative thinking, positive energy, fluid communications* and *robust collaboration.*

Managers of highly effective teams, whether traditional or distributed, are successful in creating a purposeful environment that motivates team members to self-organize productively* to make progress toward the mission objectives (Ancona and Bresman 2007). One recommendation for managers of distributed teams is to visit team members in person reasonably often,† if possible, to reinforce the principles of the organization and to build personal trust between the manager and the team.‡ To bolster the purposeful environment it helps to have continual communication and follow-up on commitments. A concerted effort to maintain personal relationships is needed to sustain high performance.

---

* This can mean that these managers help the team organize their roles and responsibilities, within which they can certainly self-organize, but by staying in their lanes of responsibility. On the other hand, a less heavy-handed management approach would be to ask team members to collectively determine their own roles and responsibilities without creating artificial partitions restricting team interactions. In the latter case, as the team self-organized, management would only step in to ensure that all aspects of the problem got coverage.

† However, visiting face-to face is not essential. Team managers deal effectively with many people regularly that they rarely meet face to face. The main point is that regular and significant interaction between the manager and team members, using whatever communication modes are available, is good.

‡ Refer to the section in this chapter titled "Implications for Leadership in Chaordic Organizations."

Regularly scheduled update meetings, using audiovisual conferencing technologies, are useful provided they are run efficiently.* They create opportunities to review recent accomplishments, share information on current events, and generate items for future action. In addition, a process that includes group meetings held on an ad hoc basis to share the most exciting developments and thoughts, and to celebrate occasionally, is recommended. If people adopt a culture of promptness, both types of meetings can be of remarkably short duration (well under an hour, and sometimes only 15 minutes, say). In face-to-face gatherings, productive, short daily meetings can be encouraged by standing rather than sitting (Smits 2007). In teleconferences, ensuring meeting setup beforehand by support staff can avoid inevitable time-wasting distraction from the business at hand, and possible disgruntlement of management and technical staff.

Information Sharing

In distributed environments it is even more important to share information in ways that build trust and enhance productivity and mission success.† This is not easy, for reasons of negative cultural and managerial biases to be explained in the following text. We feel that groups who are better at sharing information will outperform those who are not. By implication, then, teams should either be given—or take—more latitude in sharing information. If and when the advantages of this approach manifest themselves, external managers may change their ways, and the practice of

---

* Mixing meeting methods is also a good idea. For example, quarterly meetings might be held face to face, and monthly meetings via VTCs.
† Information sharing is a behavior that may be reinforced (or not) by cultural norms and influenced by leaders. Trust building may include the exchange of information, but only if the information is timely and properly set in a context. Information sharing is more than a transactional exchange; it also has symbolic valence.

information sharing will go beyond the lip service stage. Paraphrasing a statement made by a character (New York D.A.) in the 2009 movie *The International*, managers know what they want to hear, will accept what preserves the perceived *status quo*, but they do not want the truth because that implies the need to take responsibility, something to be avoided at all costs.

Too often, the prevailing cultures of many organizations block, or at least hinder, true information sharing. People are rarely rewarded for sharing and are usually punished in varying degrees. Many traditional managers and others in leadership positions still protect information that they feel gives them power. They will not view kindly individuals within their purview that deplete that resource by sharing such information. Giving others, particularly rival managers, the advantage of knowing their thinking is potentially threatening. This opens the door to unwelcome criticism that can cause consternation and distraction from the task of running their operations autonomously. Even when information is shared benevolently for the common good, there can be expectations and accompanying pressures to continue providing the resource without any compensation, e.g., external funding to sustain the service.

Information sharing, unprocessed as well as processed, is supposed to be the norm. That's curious. Much of the mantra, since 9/11 and even before, and at least officially in the DoD is to share information so that everyone can do their job better. But the same DoD, in order to block possibly dangerous free flow of information among its troops, for example, has tried to block the use of popular websites. Evidently this attempt is doomed to failure because there is evidence that Generation Y (Millennials) are using social sites that have thus far escaped the ban, e.g., Facebook, to self-organize their units (Plexus Institute 2008), despite being redirected by DoD to cease and desist so as not to contribute to additional troop casualties. It's clear that these younger servicemen have little realization that what one writes on these websites can be quite public. If, indeed, some troops are developing and sharing military strategy over an open, public source, that would be a security breach and such blocking by the DoD is quite justified.

The government sees a related and serious problem with the publication of classified diplomatic cables, as in the November 2010 Wikileaks scandal (Wikipedia 2011b; *Economist* 2010a, 2010b). One might think these leaks would make us more honest and open with each other, and that officials would declassify or at least downgrade most of this information, keeping derogative thoughts to ourselves instead of committing them to text. Unfortunately, like the articles imply, the reverse will probably happen, and we will classify these inane things at an even higher level!

Evidently, younger generations see no problem with sharing information of all kinds. It's part of their culture, but for security reasons is against the culture of most DoD—and other—government organizations. (However, refer to the earlier discussion, in the "Serious Games" section of this chapter, of the Air Force and the DoD using virtual environments.) Again, the young do not seem to understand privacy as an issue.* They think it is private, but it is not. This is much like rude behavior people engage in while driving their car. They forget people can see them because they are in their car. They think they have privacy, but don't.

For an excellent discussion of many of the topics touched upon previously, and associated trade-offs between more information sharing and less, from organizational and generational perspectives, refer to Sander (2008).

The authors advocate more information sharing as a way of building trust (refer to Chapter 3) and increasing the potential

---

* Increasingly, those at the mercy of malicious broadcasters of embarrassing (or worse) personal information on the Internet—that cannot be retrieved once it's out there—are committing suicide in desperation! (*New York Times* 2010)

for receiving information, in return. Information sharing strengthens mutual understanding of the environment, and can help better position one's own agenda. Information sharing, if done with respect, honestly and openly, accompanied by earnest dialog, can produce win-win situations that truly benefit all parties.

Oakley and Krug (2006) advocate working from the outside to the inside when contemplating what potential benefits accrue to which group. This emphasizes the overarching goal for the many first, followed by satisfying the mission of the organization, but does not ignore individual rewards as well. That last piece should be explicit to help harness people's energies in the cause.

Many people just don't believe information sharing can help them based on observing how people around them seem to lose power when giving up information, and further, sometimes get admonished for doing so. It's very difficult to change such ingrained cultures,* especially in organizations that have relatively autocratic leaders that insist on a top-down-driven command-and-control style.† Organizational cultural change cannot be accomplished when those in positions of power are unwilling to let go of their need for tight control; one almost needs to await a new generation of leaders, and even then, the culture may change too slowly to really make a difference for improving the success of the organization.‡

---

* Culture is a social construction. The positional leaders of an organization may fit with the culture or not. An outsider executive, for example, can cause fractious groups to come together in opposition to him or her. If that outsider executive is in government, he/she may have little power to fire people. In the private sector, he/she usually can. That's part of why the dynamics of sectors are different with regard to leadership and change.

† One would like to believe that this description reflects old behavior. It is undoubtedly true that exceptionally high-performance teams share information. Refer to (Chesbrough 2003) and (Moore 1996) for more thoughts.

‡ Organizations bear the imprint of the original leaders. Subsequent leaders may not fit with the organization if they are too far from the originals. Culture does change very slowly, but it does change.

Ancona and Bresnan (2007, 90–91) give an example: "when team members were silent about what they knew, they did not take on a leadership role but abdicated this responsibility. ... [the project manager], for his part, did not create the conditions needed for people to feel safe* and reveal the information that the team needed." They also point out that when "divergent political interests enter the team, those external conflicts can become internal team conflicts. This puts extreme demands on internal coordination and execution."

Guilds

Nurturing the (re)establishment of guilds within a distributed organization may be a way of encouraging the sharing of information (see preceding subsection) in a less threatening context. Guilds tend to be oriented toward specialized knowledge among relatively eclectic practitioners and not in the mainstream of an organization's business flow. Once guilds take hold and produce or improve upon a relevant body of knowledge, management might harvest this knowledge selectively to reap benefits for the organization.†

---

* Psychological safety has been studied (Edmonson 1999). It has some similarities to trust (refer to Chapter 3) in that it is earned and tested. If people know they're in a low (high) trust or unsafe (safe) environment, they'll act accordingly. The problems occur when people's expectations are not met in one direction or the other. That can create a disorienting situation. If the disorienting system persists and operates consistently, people may eventually accept it as the new norm.

† Do not misinterpret this comment in a way that suggests management is necessarily separate from the knowledge production system. In fact, management might include guild members, as well. But management does have the latitude to decide how to allocate resources by which the knowledge base could be searched and digested.

Guilds have had a spotty history over the centuries and were not widely in favor recently (Wikipedia 2008a).* However, maybe guilds can be brought back more strongly in chaordic environments. Because it is so difficult to establish new mind-sets for dealing with complex systems engineering situations, as we have discussed in Chapters 1 and 3, for example, building a guild structure to explore new ideas and help bring them to fruition through the organization is perhaps a good way to go. A few of the pros and cons concerning guilds are offered here.

- ■ **Option 1:** Pure, stand-alone, unfunded Guild, consisting of a coalition of dedicated members focused on a primary topic or family of closely related topics
  - – Pros
    - • Accumulates and builds a specialized knowledge base
    - • Establishes a reservoir of potential application practices
    - • Builds camaraderie and trust among participants
  - – Cons
    - • Saps energies for solving more pressing problems
    - • Lacks focus on delivery mechanisms benefitting the entire organization
    - • Develops *we* versus *them* attitude hurting organizational cohesion

---

* The guilds envisioned here would carry a model of professional development, certification, and continuity. They would be self-organized and self-sustaining. To the extent a guild receives resources, including personnel, infrastructure, and funding, from the organization, fine and dandy. But if these resources are insufficient to make the guild viable, voluntary individual efforts of guild members (on their own time) may make up the difference. Downsides of guilds might include: (1) excursions into "elitism," and (2) going off on tangents that serve just the personal interest of subsets of members and not the organization. If such things happen and they dominate the guild activities with lesser benefit to the organization, then other alternatives might be pursued. In any event, guilds should not be viewed as a silver bullet, nor should any other option.

▪ **Option 2:** Organizational technical center focused on a fundamental discipline such as computer and information processing, sensor technology, multimedia communication, networking, organizational change management, systems engineering, etc., leveraged by (and sponsoring) an informal community-of-interest guild aligned with the same discipline
  – Pros
    • Augments and improves specialized knowledge base
    • Develops better models for skills and processes/tools/methods delivery
    • Facilitates bench-building strength and skills cross-fertilization
  – Cons
    • Depends (at least to a greater extent) on external funding
    • Encourages indirect value delivery and nonstandard service offerings
    • Leads to potential rivalries with less specialized staff
▪ **Option 3:** Specialist group and guild; small (5–10 persons) specialist group focused on strategic planning, market and competitor analysis, marketing, proposal development, organizational branding, intellectual property reuse, tool development and production, complex systems engineering, etc., leveraged by (and sponsoring) a guild
  – Pros
    • Augments and improves specialized knowledge base
    • Focuses on indirect value development for the long term
    • Enables education of staff, especially with rotating membership
  – Cons
    • Depends (at least to a greater extent) on external/overhead funding

- Endangers processes/tools/methods development too far from main stream
- Slowest in building specialty bench strength throughout the organization

Options 2 and 3 were considered at the MITRE Corporation (a systems engineering [SE] and information technology [IT] organization managing FFRDCs that support many government organizations) when considering guilds when forming a new Advanced Systems Engineering Center (ASEC) in December 2006. Nevertheless, when the ASEC was formed, another option was selected involving a rather loose confederation of about ten individuals within one of MITRE's Centers that remained largely dedicated to the projects in which they were previously embedded. An informal affiliation of a half-dozen scattered but like-thinking individuals who were enthusiastically pushing complex systems engineering ideas continued to exist, as well. But neither of these groups achieved what would be considered the characteristics or status of a guild. Furthermore, the ASEC proved considerably less successful than expected and was dissolved in 2009.

Organizational Learning

What does the topic of organizational learning (OL) have to say regarding traditional versus distributed environments and teams? Mainly, we believe that organizations that learn well are more amenable to enlightened policies and practices that would benefit both their hierarchical and chaordic environments. Specifically, they will more readily embrace the potential strengths of diversity and self-organization and work to ingrain these concepts into the minds of their staff.

OL is not merely a collection of learnings by individuals in an organization. The core notion here is that the whole is greater than the sum of the parts. Granted, there are local experts on various subjects, tribal knowledge bases (and

perhaps guilds), and distributed expertise can be very efficient, just like basic division of labor. Learning doesn't always
flow through the organization, and some knowledge (e.g., of
the tacit variety) is hard to diffuse. A learning organization is
(presumably) skilled in learning practice (Carroll 2008).

According to Carroll (2008) OL is a process of increasing
the amount and usefulness of knowledge, that is, know-what
(factual knowledge), know-how (process knowledge), know-
who (network transactional knowledge), and know-why (systemic and cultural context of knowledge. OL competencies
also include inquiry, reframing ("double-loop learning;" Argyris
and Schon 1978), facilitation, storytelling skills, the balancing
of exploitation and exploration/curiosity, boundary-spanning
skills, tolerance for ambiguity, long-term visioning, and social
science skills.

Detweiler (2008) pointed out how tacit knowledge is an
important consideration in OL. Tacit knowledge should be
elicited and made explicit. Tacit knowledge affects our thinking and feeling (e.g., our cognitive structures), topics we consider thinking about processes we use and perceptions we
have about fitting into our organization. These factors affect
our choices and outcomes. *Therefore, tacit knowledge can
enable and/or constrain organizational learning.*

Detweiler suggested that we can leverage our tacit
knowledge to accelerate OL if we work across our habitual
areas, invite outsiders to program reviews, and participate
in communities of interest and practice. Collectively, we
can test or challenge knowledge, corroborate or confirm it,
establish it as a new basis of practice, and become a more
intelligent organization.

Emotion has a strong influence on OL whether we like to
admit it or not. Barfield (2008) showed case study evidence to
drive home this point: "historical studies that show links found
between emotion and organizational politics actually increase
the possibilities for understanding organizational learning.
Further, that the study of emotion contributes to a broader

understanding of systemic learning and, thus, is very important to strategic aspects of organizational learning." (Refer to Chapter 2.)

There is widespread acknowledgment that emotions are within the texture of organizing. This is in itself uncomfortable knowledge within organizations. That prompts organizational members to try to de-emotionalize emotions and make them seem rational Fineman (1993). For example, people within organizations have been presented as emotionally anorexic using terms such as dissatisfactions, stresses, preferences, attitudes, and interests rather than terms expressing envy, hate, shame, love, fear, and joy. This case shows the unwritten rule that it is inappropriate to bring emotions to work. However, individuals and groups continually manage and organize themselves* on the basis of their emotional responses to organizational issues as well as the basis of avoiding emotion.

Abuzaakouk and Creekmore (2008) pointed out that (Senge et al. 1990) defined five disciplines of the learning organization, personal mastery, mental models, shared vision, team learning, and systems thinking.

Team learning is of special interest here in this chapter, and of course systems thinking is key to all of complex systems engineering. Furthermore, Abuzaakouk and Creekmore (2008) highlighted that Sanchez (2005) developed *Five Learning Cycles of a Learning Organization Model* describing and prescribing how individuals and groups with alternative interpretive frameworks (current set of beliefs about cause-and-effect relationships; culture, systems, and processes) propose new beliefs (new kinds of knowledge) to help drive OL processes (an organization's interpretive frameworks) across the organization.

Interviewing your top leaders to find out what they think about OL, as well as how they learn personally, is a good approach to finding out about what might work in amplifying

---

* Emotional intelligence plays a role in determining who emerges as leaders in self-forming groups (Goleman, Boyatzis, and McKee 2002).

good OL practices within your organization. Here's a sampling of the results of interviewing seven executives and senior managers/engineers from the MITRE Corporation (Buckley 2008):

■ Challenges
  – Understanding substance of customer missions
  – Expert culture
  – Identifying actionable people and tools
  – Colocating with sponsors
  – Pushing to delivery
  – Viewing things more broadly than customer's mission
  – Creating tension between process and innovation
  – Dealing with time pressures and competing demands
  – Capturing and applying lessons learned
■ Best practices
  – Powering the team
  – Networking socially
  – Operating without barriers
  – Getting researchers in touch with customers
  – Listening, then discussing
  – Walking in other people's shoes
  – Building strong relationships
  – Using incentive/reward programs
  – Acting more than talking
  – Making lessons learned more consumable and available
■ Personal learning
  – Taking informed risks
  – Doing and learning from one's mistakes
  – Learning from anyone
  – Listening and observing
  – Walking around, getting to know people
  – Exploring multiple roles
  – Meeting with staff in ad hoc fashion
■ Taking advantage of mentors and coaches

Implications for Leadership in Chaordic Organizations

Fundamentally, it is critical that leaders become acquainted with complementary ways of doing things that are based on an understanding of complexity theory and complex systems. As mentioned in Chapter 1, a good discussion of complexity and its context in terms of a spectrum covering individual humans through human civilizations is contained in Bar-Yam (2002). In particular, see Bar-Yam's fourth figure and the associated text for a depiction of an organizational control hierarchy, a distributed network, and an intermediate hybrid that is quite relevant the present discussion.

Oakley and Krug (2006) offer a simple framework for good leadership that is embodied in five straightforward steps:

▪ Focus on the successes you are already having.
▪ Analyze those successes for what made them work.
▪ Continually clarify your goals or objectives.
▪ Determine the benefits of achieving those objectives.
▪ Establish an action plan and accountability.

The book goes on to explain briefly how this framework can be applied effectively to accomplish the following:

▪ Conflict resolution
▪ Win-win negotiation
▪ Team and individual performance improvement
▪ Forward-focused project reviews
▪ Problem solving
▪ Developing collaboration and teamwork
▪ Effective project startup
▪ Effective selling

Ancona and Bresman (2007, 41, 47) point out that in our changing organizational world "where there was once a strict hierarchy to make decisions leadership has been pushed down.

... middle management is vanishing. ... a move from a tight structure of command-and-control toward a looser organization of coordinate-and-cultivate. ... the executive level needs to create an organization that cultivates entrepreneurial activity ..."

A marketplace mentality can be helpful in distributed environments.* Provided they have an established conducive or supportive culture, more organizations ought to encourage the internal marketing for products and services and the associated resources necessary to develop them. A simple example can be cited concerning internal research and development (IR&D). Proposers of good research ideas are more likely to attract talented contributors who want to work on the projects, thereby increasing the chances of the success of not only the research proposal but also the result of the research.† These staff could have remained committed to other projects which thereby will not compete as well.

Leaders who have official positions of responsibility within a decentralized organization's hierarchy must do the following:

■ Be attuned to the organization's external environment and what's happening internally
■ Contribute to and promulgate the organization's vision to the distributed workforce
■ Ensure that people are being measured and evaluated properly, i.e., for efforts and achievements that align well with the organization's vision

---

* Some caution is warranted. If a distributed environment is merely a marketplace in which transactions occur, there will not be enough social fabric or social capital to enable organizational learning and robust teams, especially to address complex issues.
† There are several implied variables in this sentence. First, there is the intrinsic quality of the research idea, which includes the research questions, the intended methodology, and the basis in theory, and the implications for practice. Second, there is the perceived appeal of the proposer in terms of domain competence, ability to attract resources like investment, laboratory space and equipment, and the proposer's perceived effectiveness regarding the task (i.e., can he/she complete the task? Would he/she be good to work with? Would he/she be able to write up, publish, and/or patent the project findings?).

- Create conditions for purposeful collaboration in furthering the organization's mission
- Reward informed risk-taking accomplishments of self-organized groups and individuals
- Observe and learn from the impacts of distributed organizational activities
- Continually assess the level of organizational trust and foster improvements to conditions that engender increased trust
- Decide what actions are warranted and when
- Continually modify the organization's approach to its work to adapt to external events
- Lead by example with behaviors consistent with what is being advocated
- Be open and honest in all business dealings and interpersonal relationships

Other leaders, i.e., those without formal leadership or management responsibilities, should do the following:

- Internalize the vision and mission of the organization to guide their day-to-day activities
- Be aware of their potential power to impact positively the entire distributed organization through their actions in support of the vision and mission
- Be sensitive to opportunities for outreach and collaboration with others, not only within the organization, but also contacts made externally
- Proactively pursue collaborative efforts, observe and monitor those that seem to gain traction, and amplify successful endeavors while tabling those less so
- Regularly report progress (or lack thereof) to the supervisors and others in the hierarchical management chain to keep those leaders informed and unsurprised
- Freely share and comment on information gleaned from various sources that might be of interest to others

- Take some time to contribute ideas to others in response to their attempts to share information
- Adopt a professional philosophy embracing openness, honestly, integrity, healthy (objective and impersonal) debate, information sharing, adaptation, and self-organization
- Bring humility to the subject and respect the opinions of others.

Ancona and Bresman (2007, 216) say that "the organization needs to have a structure in place that provides a fertile soil for distributed leadership (Malone 2004) through X-teams. The steps top management can take to create a supportive context …" are to provide strategic direction, manage overload and empower people, be ambidextrous, promote networks, provide temporal leadership, and be role models. Although the meanings of these terms are rather self-evident and straightforward, they are elaborated upon in their book, so the reader is encouraged to delve deeper if so inclined.

Pierce and Hansen (2008) is an excellent scholarly paper that yielded practical insights about the qualities of leadership that engender trust, and therefore higher performance, within virtual teams. Here are some quoted highlights from, this paper.

> [P]ersonality traits of virtual leaders have a substantive influence on the perceived effectiveness of their teams, … [Those] who are agreeable, emotionally stable, extraverted, open, and conscientious … engender feelings of trust … assessment of leadership potential … is less about … ability to directly affect the outcomes … than the … capacity for fostering trust. …
>
> The implications for managers and other organizational leaders are substantial. … virtual teams would benefit from having their members share personal information with each other. Team building exercises that help members discover the things they have in common will help strengthen their bond and cognitive trust in each other.

Summary

Based upon the discussion of this chapter, Table 4.1 contains a (grossly simplified) characterization of traditional, distributed, and almost fully virtual team environments.* As can be appreciated, the term *team* should be used with care. *Team* may convey the wrong meaning to the constituents, particularly those ensconced in a distributed organization or the virtual environment that is expected to become more prevalent in the future. Indeed, the term team may induce a negative connotation in some who want to contribute to the organization but may not want to belong to a team *per se*.

The key to successful organizational structures or restructuring in a chaordic complex world is a clear understanding of the talents, shortcomings, and motivations of as many of the people involved as possible.† Leaders must recognize that there is no cookie-cutter solution that fits all organizations and individuals. This would be trying to follow flawed management theories of the past. Leaders must tailor techniques to the individuals and the individual groups.

One can start by addressing main organizational goals, then refining them into departmental goals, and individual goals along with structuring tasks intended to meet such

---

* A fully virtual environment implies working from home or various other locations not associated with the organization's physical plant(s), i.e., a highly variable physical infrastructure.

† This goes to issues of organization theory and practice. The physical structure of an organization can change over time. A single organization may utilize technology to support many ways of working. The culture of the organization will be the same regardless of the technology base used in a particular instance, though the members' behaviors may vary depending upon the technologies available, preferred, and routinely used. Organizing is a means to one or more ends, just as technology provides various means to support various ways of organizing. In the scholarly literature, this is the principle of *equifinality* (Wikipedia 2010), or multiple paths to a destination.

Table 4.1 Characteristics of Team Environments

Characteristic	Team Environment		
	Traditional	*Distributed*	*Virtual*
Organization	Top-down	Loosely coupled and adaptive	Self-organized
Direction	Management	Exploration	Mission
Meetings	Face-to-face	E-mail and teleconferences	Social networks
Representation of Members	Their organizations	Their organizational interests	Acknowledged inter-dependencies
Dominance	Team	Blend of team and individual	Individuals
Rules	Consensus	Influence	Marketplace of ideas
Camaraderie	Team-building events	Ad hoc personal relationships	Value-added respect
Dissonance	Not invented here	Distrust of management	Lack of focus
Productivity	Mechanistic	Opportunistic	Diverse
Survivability	Constrained	Flexible	Robust

goals. This is best done by the participants at each level of the hierarchy.*

In a group setting, look for overlap of goals—streamline by grouping individual goals and naming them. Prioritize goals or areas of focus, and drop tasks not needed to achieve agreed-to goals.

---

* The key here is that the goals need to be aligned but not rigidly coupled. People at all levels of the organization can be ahead of the official organization in figuring out what is needed, how to get there, etc. If the goals become something to which each employee must be very tightly coupled, there will be little room for innovation, creativity, and vision.

Establish fluid teams of two–three individuals who work well together to focus on projects/goals—there will be overlap due to expertise and cross-training. Add additional individuals to groups as social skills in group dynamics increase. Keep the overarching hierarchical structure to a minimum but provide guidance to overall organizational goals.

Use of fluid small teams can be enhanced by cross-training individuals to be able to join any group at any time to meet new challenges to organization. As they better understand how the organization works, individuals can be encouraged to apply their problem-solving abilities, and they will be less fearful of taking on things with which they are unfamiliar. This will increase success and decrease serious failures.

Keep individuals challenged by engaging them to feel more invested and responsible for success of organizational goals. Improve team dynamics through coaching and harnessing the ever-increasing experience and confidence of individuals. Help individuals see the Big Picture of how variables fit together.

The overall health of the organization can be evaluated by observing the flows of informal social networks and the ideas produced. Such is the essence of organizational learning.

Takeaways

There are many approaches to improving team performance. A few that we feel are worthy of highlighting are mentioned in this chapter. As stated in the Introduction to this chapter, we have not attempted a comprehensive literature search. Nevertheless, in these few examples we see reinforcement of several common principles that we espouse here as takeaways for the reader. Good progress toward organizational goals can be achieved by doing the following:

■ Establishing a vision and defining a mission that can be internalized

- Keeping overall objectives in mind, taking holistic approaches to solving problems
- Including stakeholders, and analyzing and stewarding them to the extent possible
- Paying attention to what motivates people and creating appropriate incentive structures
- Creating conditions for self-organization and facilitating interpersonal relationships
- Communicating frequently in whatever modes are available
- Sharing information in counter-to-prevailing-culture ways
- Tolerating but shaping threatening outlier behaviors in more productive directions
- Welcoming team membership changes to strengthen collective capabilities
- Being open and honest in all relationships and business dealings
- Observing what's happening in the external environment as well as internally
- Continually reassessing the situation, looking for what works and what is missing
- Pursuing enterprise opportunities and encouraging informed risk taking
- Learning from mistakes while adopting a positive outlook, not dwelling on the negatives
- Planning and replanning, adapting one's actions so as not to be wedded to a fixed path
- Rewarding and celebrating meaningful results.

Chapter 5

Application of Theory

B. G. McCarter

Complex Adaptive Systems: A Reprise of Previous Chapters

Complex adaptive systems are much like the incredible unchoreographed interplay of organisms in the vast oceans, African savannas, and arable and/or ranch land, for example. Much of what occurs under the water's surface, in the brush, or within the soil or plain is unseen by us. Yet, the interactions occur constantly, simultaneously, and on a wide variety of levels. Each organism, no matter how small or large, plays a critical role in this vast complex system.

Change is a natural part of successful evolving systems. While change may be disruptive or destructive at times, it does bring in new sources of innovation and allows the system to adapt to its new environment. Invasive species in ecosystems are good examples. They change the dynamics at play in an established ecosystem, often resulting in the extinction of native species. But even native species at one time were the invaders to that area. They helped to change the environment, as well, and caused the extinction of other species in that particular niche. However, the introduction of new species does result in innovation and more diversity in the ecosystem (Pinker 2009). Genetics research has shown that ecosystems without competition result in species that are less adaptable and flexible in changing complex environments. Those ecosystems with tremendous diversity and competition are more robust and have greater flexibility in dealing with an ever-changing world.

Similar lessons can be applied to organizations and culture. There are many sources of change for human systems, most are seen as destructive or bad, but they force a complacent system to change, which often results in healthy growth, more diversity and robustness, and increased resilience, all indications of increased complexity. The Gulf of Mexico oil crisis of 2010 is a recent example. It forced an earnest search for viable energy alternatives that will, in turn, drive new industries of economic growth—eventually (Calmes 2010; Pfeifer and White 2011).

In order to understand complex adaptive systems, one must let go of the idea that you already have to know the answer in order to solve the current problem you face. Much like viewing distant nebulae in space, or an object at night, one must use averted vision in order to see your way through complex systems. If you try to tease out specific variables within a complex system that may be causing the issue or problems that you perceive, you often find yourself going down an infinite Byzantine system of rabbit holes. As if trying to grab smoke, it slips through your fingers. Complex adaptive systems are living systems, especially if they are comprised of, or include, people.[*] As such, the variables at play can be infinite. The interactions are uncontrollable, unpredictable, and the results are not reproducible. You cannot deconstruct a complex adaptive system, e.g., by election or polling results (Gill 2011).

---

[*] Even if a complex system is comprised of only inanimate objects, the overall system can appear to be alive because of the nature of the interactions of its elements among themselves and with the system environment.

To try to solve issues that arise within a complex adaptive system, one must also let go of the idea that you are in control. Organizations, being the living systems they are, cannot be controlled. We humans will try very hard to eliminate uncertainty and ambiguity in our lives in order to control those systems with which we interact. But if we embrace uncertainty and better understand the dynamics found in the natural world of constantly evolving systems, we can find our way to navigate the seemingly confusing waters and landscapes of complexity.

The most basic fundamental aspect of people and their interactions are shared conversations and stories. These dialogues and narratives affect us from the cradle to the grave. They impact individuals, groups, organizations, villages and cities, cultures and countries, i.e., the whole world. These interactions help us see our way through the ever-increasing complexity we face; the lack of social intercourse can lead to our destruction. But how do we facilitate these communications to help groups successfully move out of chaos and work with the complexity in which they find themselves?

In the twenty-first century, the world has become more globalized largely due to the Internet's making the world a highly interactive system. We have become interconnected through our Internet communications, and competition has become keen. Organizations must respond almost instantly to global changes or risk failure. Decentralization has become the mantra of the twenty-first century as small teams of individuals are now empowered to accomplish goals or projects on their own in order to help an organization reach its overarching vision or purpose. This is changing the way organizations are structured. The traditional centralized-control (or command-and-control) hierarchical structure is giving way to a chaordic structure. The term *chaordic* (a combination of *chaos* and *order*) (Wikipedia 2011a) was coined by Dee Hock (Hock 2000) of VISA Corporation in describing small quasi-independent bands working toward an overarching corporate goal.

Chaordic Systems

Living systems thrive in a narrow band between

chaos and order ...

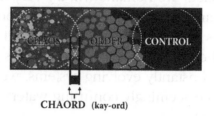

CHAORD (kay-ord)

Figure 5.1 Chaordic systems. (From Chaordic Commons, http://www. chaordic.org/definitions.html.)

This has been described as a fundamental organizing principle of evolution and nature (Hock 1999) that enables living organisms to be flexible and adaptable in their ever-changing environments. Organizations, indeed, are living chaordic systems as depicted in Figure 5.1.

Descartes was wrong when he promoted that man was a rational being, akin to a machine (Damasio 2005). And the management theories that followed were wrong to treat people in organizations as if they were machines, following the Skinnerian models of behavior management in a linear cause-and-effect fashion. As we described in Chapter 2, the way our neuro-anatomy and neuro-chemistry work makes our thinking and decision-making processes very much un-machine like. We are emotional, as well as rational, and our actions, especially those of a shorter-term nature, are driven more by our emotions. As a result, we need to understand the general principles of how people work; we need to understand people and their interactions in terms of living systems, and apply that to our management and organizational practices.

The conversations people have or do not have are often influenced by a host of personal relationship issues. The incredible diversity of individual perspectives results from our views of reality (White 2007; White and McCarter 2009). Being on the inside of a complex system and unable to have an observer's viewpoint, we cannot see the entire complex system, only a small part of it. Each individual sees a different small part influenced by our individual experiences, culture, ethnicity, race, brain chemistry, neuro-anatomy, genetics, etc. And this diversity, while it enhances creativity, is also the root of conflict in groups (Smith and Berg 1987). It is very difficult for individuals to comprehend that not only do other people have different views of reality from their own, but also that these differing views are all small parts or truths of the complete complex fabric in which we reside.

These differing views of reality affect trust within and among groups. And as we saw in Chapter 3, trust is key to the conversations decentralized organizations must have to succeed globally today. But how do you build trust and facilitate groups working together for a common goal? How do you empower management to navigate the difficult waters of trust within small groups in a decentralized world so they can actually accomplish the work required? How do you make cultural transformations happen?

Many people and organizations today feel that they are in chaos. As Richard Knowles (Knowles 2002) points out, chaos can occur when we insist on maintaining current circumstances even when it is no longer working. Our beliefs in how things should be do not fit with how things actually are, which causes a feeling that things are breaking apart. It is a prime stimulus for change, and change is very painful, both emotionally and physically. It can affect the growth of neurons and new neural pathways in our brain, which can be uncomfortable. It usually takes a dramatic event to cause us to look at things from a different perspective that enables change (Klein 2008). While chaos is uncomfortable, it serves a purpose in helping us to move within and through complex systems.

How do people in the midst of organizational chaos deal with these issues? To begin with, we have to change the way we think about organizational processes. As already noted, organizations are not machines, they are living systems. As such we have to start the needed conversations among the individuals involved in order to begin moving out of chaos and to move the organization forward though complexity to achieve a semblance of order on the edge of chaos (Wikipedia 2011b).* These conversations will help clarify our identity, strengthen and build our relationships, and enable us to gather and share information which, in turn, continues to redefine our identities and strengthen our relationships. The resulting learning makes the organization's individuals more flexible and adaptable to their environment, as a living system ought to be. Without this vital triad (sharing information, redefining identities, and strengthening relationships) functioning well, an organization can become dysfunctional.

There are many methods readily available today to help organizations move through chaos. The key, as noted in Chapter 4, is to understand the psychology of the individuals and group dynamics involved. Methods that utilize group facilitation techniques, coupled with an understanding of complexity, individual and group psychology, and general organizational principles, will do well to help an organization perform in a healthy and successful fashion.

Richard Knowles's description of the Process Enneagram (Knowles 2002, 27) is one such method that seems to be the missing link between complexity theory and practical application. It helps to identify issues in a complex system that may be causing problems in the organization. The process facilitates a shared identity among the group members in the organization, helps to build an understanding of differing views of reality, enhances communication among group members, and helps to build that most important element: trust. The Process Enneagram also helps identify problems lying hidden or undiscussed under the surface.

---

* Also refer to Figure 1.1 of Chapter 1.

In order to work with individuals in a group setting like the Process Enneagram, it is helpful to understand how our brains work, how we have come to have such incredible diversity and differing views of reality. One needs to understand the interpreter part of our brain and its role in telling ourselves stories to help us make sense of the things in our lives that we don't understand (Gazzaniga 2007). It enables us to give ourselves a feeling of certainty in life, that we have predictability. Ambiguity is seemingly banished from our lives. This can become our view of reality to help us make sense of our individual world and to keep moving forward in it with purpose and certainty. These views of reality, however, can become unshakable, inflexible, and narrow. And when that happens, it can dramatically affect individual and group interactions in organizations. If you are facilitating a group process, you need to be aware of these processes and how to handle them.

Our historical way of thinking about organizations is not working today. A paradigm shift is needed. The traditional hierarchical command-and-control method is an oxymoron in today's complex world. One cannot control complexity; by its very definition that is impossible. And because the machine image is not relevant, neither are tool kits in gathering methods to help facilitate people and their interactions in organizations. One uses tools with machines, and continuing to use such labels or language continues an inaccurate perception of an organization as a machine. Organizations are living systems, just as the people who comprise them are. Language and the words we use can affect the way we think and the way we look at problems as we search for solutions. The way we think, in turn, can affect the way we behave. A paradigm shift in the way we think about organizations is critical if we are to move forward in solving the problems of complexity facing them today. How we use language, the actual terms or words we use, can, in turn, form or shape our conceptions and beliefs.

As we look at chaordic systems and decentralized groups it becomes apparent that empowering others to do their work brings more power back to those in management positions (Greene 2000). It helps the organization to move forward and to become flexible and adaptable. The organization becomes stronger and healthier when such a system is encouraged and facilitated.

This paradigm shift also refers to changing the way individuals relate to one another. Through the shared conversations encouraged by the Process Enneagram, we can facilitate the expansion of individuals' rigidity of thought boundaries. When people begin to realize through shared conversations and developing trust that others have different views of reality and that all the varied views make up a more complete picture of the whole complex system in which we all reside, then a greater understanding and acceptance of others can begin. We learn to embrace and accept the uncertainty and ambiguity that is part of the life of living systems. Complexity is all about uncertainty. This acceptance frees us from fear and allows us to look forward to embrace new opportunities (Rebovich and White 2011; Chapter 5) as they arise and to see multiple perspectives of problems facing us. That, in turn, leads to solutions that otherwise would never have been imagined.

Conflict

Interpersonal conflict is a natural consequence of individuals having different views of reality. While many organizations dislike conflict and work hard to sweep it under the rug in favor of superficial group acquiescence, conflict in its healthy and constructive form should be embraced instead. It is through conflict that creativity is born. And to solve the enormous challenges in today's world, creativity needs to be encouraged. Healthy conflict acknowledges many different views of reality and seeks to allow individual voices to be expressed and heard. Different ideas are encouraged to be expressed, even passionately. Arguing about ideas is encouraged, but attacking others personally for their ideas is not. That is when conflict can become harmful and destructive: attack ideas, not people. Embracing conflict in groups allows all views to be heard and helps the best decision of the group to emerge and be accepted by all members. Through this process you can gain true commitment from the group, not lip service. Healthy, constructive conflict minimizes the us-versus-them mentality that accompanies most group interactions by helping people to see other's views of reality and to expand their own.

Paradoxically, decentralization in organizations can increase distrust. The smaller, physically separated groups are often unrelated to each other and do not have a shared, collective sense of identity. Not only do individuals have different views of reality, but so do groups. Each of these groups also may have their own culture and history, which may conflict with other groups within an organization. It can become difficult to confirm the truth of the stories others tell in order to determine their trustworthiness (Smith and Berg 1987). Decentralization can encourage us-versus-them dichotomies, which are the most basic element in individual and/or group conflict. Organizational collapse can be exacerbated by decentralization because of this mentality. This makes it even more apparent that the conversations among members of groups and among groups are critical in helping to build trust throughout the organization.

The various methods in use today to help solidify cultural transformation all utilize some form of basic group/individual counseling techniques. These techniques are built upon a history in counseling psychology that understands how people think, feel, and relate to one another. These techniques help individuals and groups move through change. A proper application of group dynamics techniques in organizations is the key to moving organizations forward through complex problems. The various methods being used today help people inside complex systems step outside their normal views of reality and help them to see problems, solutions, and people from a different perspective. These methods also help people to see that the complexity surrounding them is a dynamic living system, and that we are inside that complex system and are a part of it.

To dance on the edge of chaos, to maximize creativity and potential, and not go over the edge into a chaotic form of collapse requires that information flows freely among individuals and groups. Interdependence of individuals and relationships grows as a result of this information flow. All parts of the system are valued, and clarity about who we are and what we are supposed to do in an organization gains strength. This in turn increases the understanding of what the individuals within a group or organization need to do together to help the organization move forward. As individuals co-create their futures, they build trust and help gain meaning in their work (Knowles 2002, 86). In complex systems, everything and everyone is interrelated to some degree. Even small changes have huge effects within the system.

Process Enneagram©

According to Knowles and the Process Enneagram© there are three conditions necessary for small groups to be able to self-organize, which is a basic principle of complex systems. These include the free flow of information, the ability to build relationships and interdependence (trust), and have a shared sense of identity within the organization.

In helping groups and facilitating their particular complex systems problem solving it is important to understand that the answers lie within the people of the organization. They do not lie outside the system. The solutions will arise as the group gains greater knowledge, understanding, and new perspectives as they grow and evolve through the process of the shared conversation. It is also important to remember that we do find our sense of self through our relationships with others (Smith and Berg 1987). Our connectedness allows information to flow freely. It is important that conversations are encouraged to be honest, constructive, and in a healthy format. Equally important is the understanding that change is always present and adaptability is always necessary in complex systems. Nothing ever stays the same in the natural world. If one does not change, then one no longer is able to function well. This means we must embrace uncertainty and accept ambiguity, which are constants of natural systems.

So what are some of the basic principles for facilitating the conversation among individuals and groups? Richard Knowles (2002) lays out the basic structure of the Process Enneagram in the following parts:

■ The Bowl, or boundary that defines the organization by its vision and goals
■ Understanding the six key aspects of organizations as living systems
■ Understanding stages of change within organizations
■ Nine points or elements of the Process Enneagram
■ Characteristics of dysfunctional models of organizations vs. functional ones
■ Domains of the self-organizing system in organizations
■ Various simultaneous patterns of living systems in organizations

Like all complex systems, the various parts of the system are interacting simultaneously at all times. No part of a complex system is in isolation independent of the interactions occurring within the rest the complex system.

The Bowl, as described by Knowles, defines the environment in which a complex system operates. In order to problem solve within a complex system, that part of the system on which you want to focus needs to be defined. If the problem or complex system is not bounded or defined, it becomes almost impossible to constructively deal with it. Organizations are bounded by Knowles's Bowl. Here the organization's vision and goals define its particular complex system. The multitude of variables operating within the organization is then able to direct their behaviors and interactions to fall within these prescribed boundaries. Knowles discusses the various characteristics of the Bowl as including things such as the following:

- Shared identity within the organization
- Sense of knowing what those within the organization are supposed to be doing
- Shared values
- Understanding of issues at hand
- Developing relationships among members of the organization
- Understanding the surrounding competitive environment in which organization finds itself
- Keeping organizational intention and work aligned
- Making sure that information is easily obtained and transferred within the organization
- Making sure members of the organization are constantly able to learn and evolve within the organization.

The overarching question that bounds the conversation involving the following nine variables, shown in Figure 5.2, establishes the Bowl: What is it you want to solve or do?

0/9 Identity
  1. Intention
  2. Issues
  3. Relationships
  4. Principles and Standards
  5. The Work
  6. Information

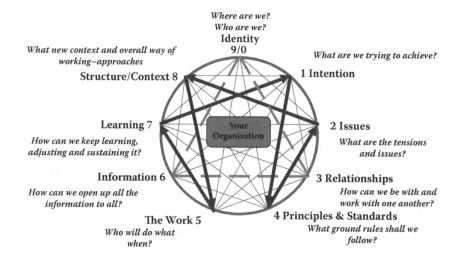

Figure 5.2   The Enneagram. (From Richard N. Knowles. *The Leadership Dance: Pathways to Extraordinary Organizational Effectiveness*, 3rd ed. Niagara Falls, NY: The Center for Self-Organizing Leadership, 2002, 35, 39.)

7. Learning

8. Structures/Context

As Knowles points out, the structure provided by the Bowl in an organization provides order for those working within the organization. In turn, this allows the individuals to focus on the work they individually and collectively need to do in order to move an organization forward. This provides a tremendous amount of freedom for the individuals working within the organization, which helps to unleash their creativity. Enabling members of an organization to freely share information and develop relationships so that they can effectively do their work and impact the organization, gives members of an organiza-tion a sense of meaning in their work. Feeling that what we do in our work matters is a tremendous motivating force for most people (Knowles 2002, 99).

As a living system, Knowles points out that there are six key aspects to organizations. These aspects in an organization help determine whether or not the living system thrives or is crippled.

- One aspect includes the kind of structures in place and their organization. If these structures are more fluid and dynamic, such as in the case of task force or self-organizing teams, the next evolution or change within the company will be much more smoothly and easily traversed.
- Flexibility in the roles assumed by individuals in the organization can further enhance the overall flexibility of the organization and its ability to adapt to changing the environment.
- A willingness to embrace change by members of an organization can enhance the organization's ability to react in a timely manner to changes experienced in the environment.
- Another key aspect is how members of an organization determine what they may want to change.
- Do members of the organization consciously learn from their environment and adapt? Do they learn from their mistakes? Do they engage in self-reflection? Living systems that constantly evolve and adapt to their ever-changing environment do just that.
- Finally, Knowles lists a sixth aspect that includes embracing uncertainty. Understanding that one cannot control nor predict the future encourages individuals to develop contingency and what-if plans. Complex systems evolve in unpredictable and uncontrollable ways. But having a general understanding of how the system reacts one can come up with a series of possible scenarios to use as a system evolves and reveals itself (Knowles 2002, 140–142).

Change

Change is a critical component of all self-organizing, evolving, and adaptable systems. But change, even the prospect of change, is often uncomfortable and resisted by people.

Knowles reflects on the various stages or rhythms of change experienced by individuals and organizations. These include flowing, staccato, chaos, lyric, and stillness. In the flowing stage, one freely engages with information as it comes to us. The flow of information confirms our beliefs in how things should work. We are not challenged; the things make sense to us. Soon, however, we enter the staccato phase tightly holding onto what is familiar despite the fact all indications are that we need change. We don't understand what is wrong but we are feeling disruptions that challenge what we believe. Next, one enters the chaos phase. The disruptions are coming fast and furious at this point; it is obvious something is wrong. We feel that things are out of control and are breaking apart. Still insisting on what we've always believed to be the way to make things work, we try to slam a square peg into a round hole and make it fit.

The key at this point is to understand that change is needed. A paradigm shift in the way one thinks. A new perspective is needed, and we need to look for new possibilities. At last the lyric phase is entered where we are able to see the changes that have occurred all around us. We become aware of the emergent properties of the complex system in which we reside. As such, it is important that individuals of the organization self-organize at this point, sharing information and building their relationships in order to evolve and meet the changes that have occurred. Finally, an organization enters the stillness space. Here we see the changes that have occurred in the environment, we become conscious of all we as individuals of an organization need to do in order to meet those changes, and we have obtained greater understanding of our role in this constantly changing complex system (Knowles 2002, 128).

How to Facilitate Change in Organizations

The Process Enneagram itself helps members of an organization identify the changes necessary in a flexible and adaptable organization. It is also a continuous process that includes a feedback loop for learning from new information and experimentation to see what works and what may not. There are nine points (refer to Figure 5.2) that are examined in the process described by Knowles (2002, 28, 29):

- **Identity:** Who are the members of an organization's history, individually and collectively?
- **Intention:** What is it the members of the organization are trying to do?
- **Issues:** What problems face organizational members? What dilemmas, paradoxes, and questions do thay have?
- **Relationships:** What is the level of trust among the members of the group in the organization, and how much do they trust the organization? To what extent do they readily interact openly and honestly with one another?
- **Principles and standards:** What are the real, often undiscoverable, ground rules for the organization?
- **Work:** What is it they actually do?
- **Information:** How do members of the organization create and share information?
- **Learning:** To what extent do members of the organization learn from their mistakes; are they conscious of what happens in an organization, and reflect on the implications?
- **Structure and context:** What is the formalized organizational structure? How are members of the organization organized? What is the environment in which they are working?

■ **New identity:** As members of an organization go through the process, sharing information and learning, they grow and evolve and gain greater insight into who they are within the organization

These nine Enneagram points represent dynamic variables that are simultaneously interacting on a constant basis within an organization. These dynamic variables are nonlinear. They do not follow a lockstep pattern. Instead, like all complex systems, they are all interacting continuously with each other making up the living system of the organization. The Process Enneagram allows individuals to unravel the seeming chaos in which they find themselves in an organization undergoing change. Working in a group consisting of key players in an organization needing to develop and direct change, a facilitator leads them through the Process Enneagram examining each of the variables listed, seeing how they impact one another, identifying possible difficulties, and determining courses of action to take as a group, being mindful of the processes and how they relate to one another.

According to Knowles, there are three basic layers:

■ Command-and-control
■ Work
■ Self-organization

that operate simultaneously in all healthy organizations. These three layers, shown in Figure 5.3, act in tandem to help create an organization's complexity. As noted in the figure, during times of crisis, a highly functioning team is able to dispense with the command-and-control process and concentrate on the *work* within their *self-organization* domain. Usually, the urgency of the crisis dictates that there is no time to modify the established command-and-control process.

The top layer is the command-and-control process that, when properly applied, establishes the general Bowl that

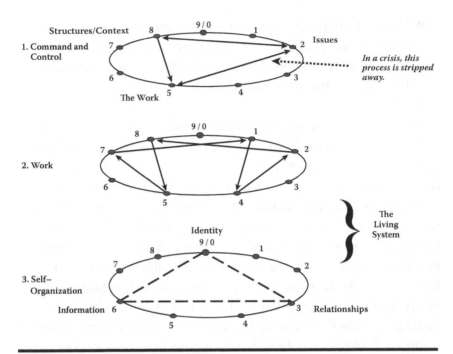

Figure 5.3 Iterative processes of living systems. (From Richard N. Knowles. *The Leadership Dance: Pathways to Extraordinary Organizational Effectiveness*, 3rd ed. Niagara Falls, NY: The Center for Self-Organizing Leadership, 2002, 38.)

gives structure to the organization and allows employees to freely self-organize underneath its overarching umbrella. Here, however, the command-and-control structure must not be too rigid as management establishes a structure and possible solutions for the identified issues, as well as which work is necessary. Together, the Work Layer and the Self-Organization Layer make up the living system of the organization.

Many organizations follow the traditional command-and-control hierarchical structure. Too heavy an emphasis on this style of organization can lead to dysfunction. Knowles expounds at length on characteristics that epitomize this type of organization. Basically, management defines goals, identifies problems, and issues orders for the workers to follow to solve them. Overall, management infrastructure tries to be rational and objective at all times. Emotions are usually treated with

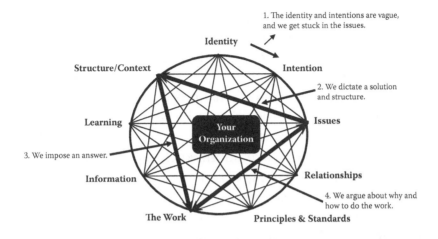

Figure 5.4 Command-and-control pattern and process (bold triangle). (From Richard N. Knowles. *The Leadership Dance: Pathways to Extraordinary Organizational Effectiveness*, 3rd ed. Niagara Falls, NY: The Center for Self-Organizing Leadership, 2002, 32.)

suspicion. Plans to try to help the organizational processes move forward are usually referred to as *management tools*. Managers are often looking to find new procedures to put in their toolkits. But as emphasized before, man is not a machine nor is the organization, which is made up of people. People are often irrational and emotional. As a result, in this age of complexity, toolkits do not apply, and have not often worked completely well. Using the command-and-control structure highlighted by the heavy (bold) triangle in Figure 5.4, managers try to control the environment and the task. However, the problem is that one cannot control complexity or the people within a complex system.

The more managers try to control their environment and the people within it, the more chaotic the situation appears. These managers end up being seen as defensive and rigid, and the organization's effectiveness is greatly reduced. Inconsistencies, which they deny, begin to be seen in their messages to the people in their organization. Soon the ambiguities and inconsistencies become un-discussable, paralyzing the living system of the organization. If the conversation and sharing of information among the employees of an organization is blocked, then the organization becomes dysfunctional and is unable to move forward. The employees no longer trust management or the organizational processes.

In contrast, Knowles describes a healthy, self-organizing living system as one that feels that information is valid and where employees can make informed choices. Employees become committed to the goals and visions of the organization and feel free to test, within the system, ideas and theories to solve the problems that they face. They feel engaged, motivated, and respected. The conversation of the living system is encouraged in a healthy organization (Knowles 2002, 42, 43).

The second layer is called the Work Layer, which describes a healthy way for organization to unravel the seemingly chaotic environment in which they find themselves during times of change.

The order in which the work is pursued is important, otherwise order is not restored and chaos continues to rule. The recommended process for self-organization is shown in Figure 5.5. Note the proper order of addressing the nine variables (1, 4, 2, 8, 5, 7, 1, 4, …) as indicated by the heavy (bold) arrows.

First, the group needs to look at what their intention is, what they are trying to do as a group. Next they need to examine the principles and standards of their organization. What are the real, un-discussable ground rules of their organization? Next, they need to identify the issues they are facing. After that, it's time to look at the organization's structure and context. How are they organized and what's happening in the environment in which they find themselves? With all the previous issues identified, it's now time to look at the actual physical work they are doing. Finally, the group identifies new things that they have learned in the process and examine what their future potential may be. This may in turn modify their intention as a group, helping the group to evolve and adapt in a spiraling pattern as they continue the self-reflective process.

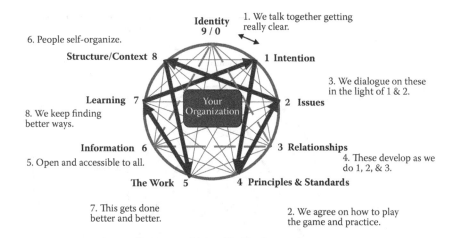

6. People self-organize.

Structure/Context 8

Learning 7
8. We keep finding
better ways.

Information 6
5. Open and accessible to all.

The Work 5

7. This gets done
better and better.

Identity
9 / 0

1. We talk together getting
really clear.

1 Intention

3. We dialogue on these
in the light of 1 & 2.

2 Issues

3 Relationships
4. These develop as we
do 1, 2, & 3.

4 Principles & Standards

2. We agree on how to play
the game and practice.

Figure 5.5   Process of self-organization. (From Richard N. Knowles.
The Leadership Dance: Pathways to Extraordinary Organizational
Effectiveness, **3rd ed. Niagara Falls, NY: The Center for Self-**
Organizing Leadership, 2002, 35.)

Knowles describes the main domains of the self-organizing
system in organizations as identity, relationships, and information. The Identity, Relationships, and Information variables are
shown by the heavy (bold) triangle in Figure 5.6. This is where
the conversations, the sense of self, the sharing of information,
and the building of relationships occurs. In times of crisis, this
dynamic leads the way in moving the mountains necessary to
restore order. As the name implies, self-organizing systems are
spontaneous. They are the informal social networks that develop
under the surface of an organization to get the work done.

Who we are as an individual or as an employee in an
organization is influenced and partly defined by the relationships we develop with others. The relationships we have with
others are established and enhanced through the sharing of
information, i.e., the conversations we have with each other.
In organizations where these conversations are stymied, especially through overly rigid command-and-control structures,
people no longer know who they are and what their roles are
in relation to the organization. They mistrust each other and

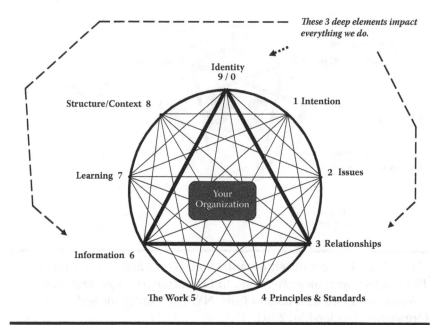

Figure 5.6   The domains of self-organization (bold triangle). (From Richard N. Knowles. *The Leadership Dance: Pathways to Extraordinary Organizational Effectiveness,* **3rd ed. Niagara Falls, NY: The Center for Self-Organizing Leadership, 2002, 33.)**

do not share information. If they do not share information, they cannot get the work done that is necessary to accomplish their stated goals.

The results of an actual application of the enneagram are depicted in Figure 5.7. A generic view of the iterative learning cycle process is shown in Figure 5.8.

Counseling Skills and Techniques

As previously stated, management tools and toolkits are not as effective in working with people as are traditional counseling skills and techniques. Counseling psychology was built on the understanding of how individuals and groups work. It was also designed to develop the techniques necessary to help people understand themselves and each other in

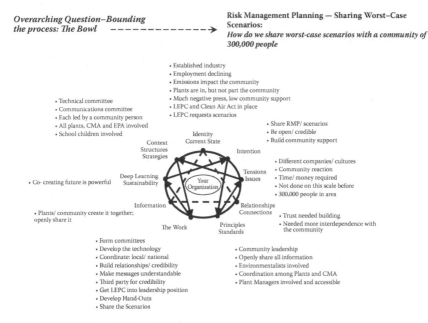

Overarching Question–Bounding the process: The Bowl ------------→ Risk Management Planning — Sharing Worst–Case Scenarios: *How do we share worst-case scenarios with a community of 300,000 people*

- Established industry
- Employment declining
- Emissions impact the community
- Plants are in, but not part the community
- Much negative press, low community support
- LEPC and Clean Air Act in place
- LEPC requests scenarios

- Technical committee
- Communications committee
- Each led by a community person
- All plants, CMA and EPA involved
- School children involved

- Share RMP/ scenarios
- Be open/ credible
- Build community support

Identity
Current State

Context
Structures
Strategies

Intention

- Different companies/ cultures
- Community reaction
- Time/ money required
- Not done on this scale before
- 300,000 people in area

Tensions
Issues

Deep Learning
Sustainability

- Co- creating future is powerful

Your
Organization

Information

- Plants/ community create it together;
openly share it

Relationships
Connections

The Work

Principles
Standards

- Trust needed building
- Needed more interdependence with
the community

- Form committees
- Develop the technology
- Coordinate: local/ national
- Build relationships/ credibility
- Make messages understandable
- Third party for credibility
- Get LEPC into leadership position
- Develop Hand-Outs
- Share the Scenarios

- Community leadership
- Openly share all information
- Environmentalists involved
- Coordination among Plants and CMA
- Plant Managers involved and accessible

Figure 5.7 Sample of documented Process Enneagram. (From Richard N. Knowles. *The Leadership Dance: Pathways to Extraordinary Organizational Effectiveness*, **3rd ed. Niagara Falls, NY: The Center for Self-Organizing Leadership, 2002, 51.)**

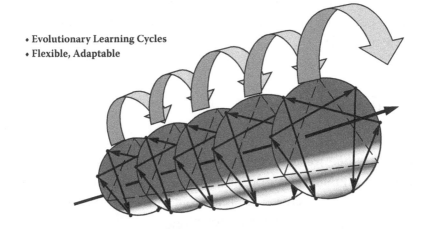

- **Evolutionary Learning Cycles**
- **Flexible, Adaptable**

Figure 5.8 Iterative learning cycles. (From Richard N. Knowles. *The Leadership Dance: Pathways to Extraordinary Organizational Effectiveness*, **3rd ed. Niagara Falls, NY: The Center for Self-Organizing Leadership, 2002, 123.)**

groups. Counseling psychology is a field that strives to understand how people come to be who they are and how that impacts their relationships with others in their environments. Understanding complexity in organizations requires that one understand the people that make up the living system of the organization. They do not respond well to traditional management processes that were designed for machines. And generally, people respond best to facilitative counseling techniques.

Once one understands the process, it is time to make sure you understand basic group dynamics and facilitation skills to help the process be successful. Here individual and group counseling techniques, which are often found in executive coaching skills (Flaherty 2010), are most useful.

It is important to understand that people are not wholly rational. We are emotional beings and are often irrational. Current research is revealing that our emotions are important in our ability to make good decisions, as opposed to the traditional belief that they hinder decision making. Our emotions cannot be separated from who we are and how we think. So, in dealing with people and groups, it must always be kept in mind that people (and the groups comprised of them) are irrational and emotional, and not machines (Lencioni 2002).

Specific Techniques

In order to change the way we interact with other people, we have to change the way we think. One of the best ways to effect that change is to not go on autopilot. That means being aware of what we say, as well as what other people are saying, and not just automatically reacting.

In order to effectively use counseling techniques to facilitate group dynamics, one needs to examine one's values. Do you respect other peoples' viewpoints? Are you willing to listen and try to understand others' viewpoints (Schwarz and

Davidson 2005)? As mentioned previously, because we are all inside a complex system and there is no external observer, all viewpoints are valid.

Core Values

According to Schwarz and Davidson (2005), there are four explicit core values involved in all facilitation processes. These include (1) valid information—sharing all relevant information you have about an issue so that others in the group understand your reasoning and why you hold the position or ideas you do; transparency of personal interest is beneficial; (2) free and informed choice—group members are able to make sound, well-derived decisions when all information concerning an issue related to all parties of the group are presented and are transparent; subtle as well as overt pressure from other parties is removed; (3) internal commitment—because all ideas of the group members are allowed to be freely expressed, including passionate expressions, individuals feel their ideas are heard and that subsequent decisions made by the group are valid; this is like dust swirling to support those decisions and commit to them; and (4) compassion—the Golden Rule is in effect here: objectively listen to what others have to say, suspending during that time period any personal judgments. Just as you would want others to objectively hear what you have to say, so too should you afford the same courtesy to others (Schwarz and Davidson 2005, 6). These core values reflect a basic respect for the opinions of others, which facilitates the communication process and builds trust.

Ground Rules

Schwarz and Davidson (2005) list nine ground rules for effectively facilitating group dynamics, which overlap the basic structural and facilitation rules for the Process Enneagram presented by Richard Knowles (2002). Both incorporate basic counseling psychology techniques. These ground rules are as follows (Schwarz and Davidson 2005, 8):

1. Test assumptions and inferences.
2. Share all relevant information.
3. Use specific examples and agree on what important words mean.
4. Explain your reasoning and intent.
5. Focus on interests, not positions.
6. Combine advocacy and inquiry.
7. Jointly design next steps and ways to test disagreements.
8. Discuss un-discussable issues.
9. Use a decision-making rule that generates a level of commitment needed.

These ground rules are geared to create transparency in groups. This means there are no hidden agendas or unknown motivations for various positions held by group members. Oftentimes misunderstandings in groups occur because we don't fully understand why something is important to someone else within the group. Understanding various individuals' needs or interests helps the group to come to a decision that incorporates those various needs. Asking nonjudgmental questions about other group members' statements or positions in order to objectively more fully understand their needs helps make sure everyone's view of reality and interests are heard. In addition, not being afraid to discuss issues that have been sources of contention and conflict for the group (the un-discussable issues) helps to clear up misunderstandings and provide greater transparency for the group process. Finally, understanding that a diversity of viewpoints can create conflict, and that conflict is not necessarily negative, frees the group to fully explore all opinions and viewpoints that are held by the group members. This, in turn, helps to facilitate individuals' commitments to the decisions that are made by the group about the various issues at hand.

Evaluating Group Processes

In order to gain individual commitment to decisions made by the group, it is important for people to feel that their opinions and feelings have been heard and considered by the group.

One of the most difficult aspects of facilitating a group process is being able to objectively recognize the group dynamics at play in real time. Being a part of the group and listening to the conversation, as well as participating in it, can make it difficult to also observe and analyze what those dynamics mean.

One of the keys to facilitating healthy group dynamics is to ensure transparency. And the best way to ensure transparency is to be open, honest, and nonjudgmental in your communication with others in the group, as well as about your interest concerning the issue at hand. Helping to keep the group communication from being an emotional reactive process requires the ability to be in the here and now during the group process. In the counseling psychology facilitation process this involves observing a behavior or statement made by an individual in the group, thinking about what that statement might mean, reflect back to the individual their statement to clarify and make sure that you heard it correctly, and if so, then share with the group what you thought that statement meant—often referring indirectly to unexpressed concerns of the individual.

It is important to note at this point to be careful not to trigger defensive reactions from the group members. The best way to do that is by taking small steps in your inferences to increase your likelihood of being accurate, and in being respectful in the way you express your inferences. During this reflective or feedback process you will check to see if the individuals agree with your inference of what was meant by the statement. It is important to note here that the reflective process is nonjudgmental and nonaccusatory. It is a respectful and compassionate exploration of individuals' ideas, concerns, and feelings about issues with which the group is wrestling (Schwarz and Davidson 2005).

The best way to help avoid receiving defensive reactions from others is by using what is known as *I talk*. If, when in conversation you refer to the other person, their statements, or behaviors, as "You said ..." or "You did ...," you have a greater likelihood of eliciting a very negative and defensive response. However, if your inferences are phrased in terms of "I feel ...," "I heard ...," or "I think ...," then you're more likely to have the other person hear and consider what you're saying. It is alright to be passionate in expressing our ideas, but to facilitate communication and understanding within a group we need to make sure that we aren't accidentally making people defensive. When people are defensive, they shut down and no longer hear what others have to say.

It is difficult to break old habits learned in our past. When we were young, we learned group behavior from our immediate family dynamics. We often learn to not express openly if we want something done in order to try to manipulate or move others to agree to our positions. As we continue to grow and mature, we added to our patterns of group behaviors from our associations with friends and classmates. Oftentimes, these dynamics do not work well in the adult world. Other adults can often feel or recognize when they're being manipulated. That can lead to resentment and mistrust in group dynamics. But changing old ways of behaving takes a conscious effort. The neuronal pathways created in our brain by our experiences when we were younger literally have to be changed. One of the reasons people resist change so adamantly is because it is difficult and painful. It is easier to follow the well-worn path of behavior patterns established over our lifetime, even if it does not work well (Philips 2006). One of the most effective ways to change our old patterns of behavior and to embrace new and healthier ones is to be consciously aware of our motivations, our statements, and what we actually do.

Knowles's Process Enneagram structure enables group members to efficiently analyze the complex variables at play in their situation while they practice healthy group patterns of interaction: facilitating the conversation, speaking in a safe space, demonstrating respect and compassion for one another, building relationships, developing trust, gaining meaning in their organizational roles, and establishing action plans to enable their organization to be more flexible and adaptable in an ever-changing world. Having a general structure by which to untangle the knot of complexity frees up some energy to focus on developing healthy group interaction dynamics. It requires constant vigilance in the beginning while you develop these new and effective techniques. The more these skills are practiced, the easier they become, and less conscious thought is required. These new and healthier ways of interacting will become automatic.

Social Systems

People and their interactions make up what is referred to as *Human Systems*. The dynamic, fluid, and ever-changing interactions among people help to create the complexity. Group behavior is often unexpected and unpredictable because the interaction of various individuals creates something new: the group. Understanding that there are both functional and dysfunctional dynamics at play within groups is important in helping groups work together and enable their organizations to be more flexible and adaptable in an ever-changing environment. These dynamics are impacted by many aspects of the individuals who make up the group as a whole. In addition to individual personalities, cultural backgrounds, life experiences, and brain chemistry, you also need to consider their experiences and attitudes concerning their work environment.

Oftentimes in traditional command-and-control hierarchal organizations, individuals are isolated from the main vision and goal of the organization. They only know their small role in the organization and the job they're expected to do. They do not necessarily understand or know how their job fits in with the overall vision of the organization. Groups work best together when their goals are clear and everyone knows not only what they are supposed to do but why they're supposed to do it. When everyone understands how their individual roles fit together to help the organization achieve its vision and goals, there is less confusion and individuals derive greater meaning and purpose in their work.

Motivation is another issue that affects how well individuals and groups work together. Much has been written historically in an effort to try to understand what best motivates individuals in the workplace. Early experiments showed that assembly line workers improved their performance not because of changes in lighting in their work environment, but because somebody was paying attention to them. Recognition for work done continues to be an effective motivation tool, as is monetary compensation. But perhaps the most effective motivator of individuals in any environment is a sense of purpose, urgency, and meaning in their work as well as in their lives.

Working collaboratively with others in the workplace as well as understanding the goals of the organization and how one's particular role impacts those goals helps to create meaning and a sense of purpose for individuals with an organization. This is a tremendous motivator for most people. Traditionally, conflict in groups has been considered to be negative and therefore to be avoided. Oftentimes this results in issues being avoided and not discussed. Un-discussed conflict in group situations results in dysfunctional dynamics. Communication can be hindered, trust can be damaged, and obstacles to achieving goals are increased. Understanding that conflict is derived from diversity and that this diversity of individuals increases creativity and the ability to problem solve as a group can lead to a more open and respectful communication among group members.

Conflict is often avoided because of the false assumption that acknowledging and dealing with conflict would hinder the group achieving consensus. In reality, avoiding openly discussing and handling conflict among group members often creates unnecessary stress within a group and ultimately interferes with the group's ability to achieve its goals. Embracing conflict in a constructive and positive manner can encourage and facilitate tremendous creativity within a group. Some individuals in a group may not work well together for a variety of reasons, but openly discussing the issue in a positive manner can find ways to deal with it. Constructively embracing conflict can also help expand individuals' views of reality. They can begin to see that there are multiple valid perspectives to almost any complex situation. In addition, this embracing of individual differences in perspectives can lead to conversations where unseen or hidden issues can be brought out and considered, as well as tapping into previously unknown and unique talents that individuals of the group may possess.

All of these group facilitation techniques help groups to self-organize better. It affords an opportunity for any individual in the group to experience leadership roles at varying times while the group is working on a project (Knowles 2002; Schwarz and Davidson 2005). Every individual has an opportunity to practice facilitating open and healthy group dynamics, which in turn not only helps the group be more flexible and adaptable, but it also helps the individual in their personal growth. As individuals of the group practice and use constructive and facilitative communication skills, they build their relationships with one another through the ensuing conversation. The relationships they develop, in turn, helped to build trust among the members of the group as well as for the group as a whole.

This gets back to what was mentioned previously about how the group is a living system according to

Richard Knowles (2002). A clear sense of the individuals' and groups' goals combined with respectful and open communication and information sharing with members of the group develop a self-organizing entity that is flexible and adaptable. Developing and maintaining trust among individuals within a group and an organization is critical for organizations to be able to afford and achieve their goals. In order to build and maintain trust, consistency in behavior is required. This goes back, again, to demonstrating by our behaviors the core values we espouse.

Self-organizing groups can feel more empowered than individuals that normally exist in a rigid command-and-control hierarchal structure. They are not isolated, they share information, they share expertise and collaborate, all of which enables the individuals of the group to accomplish more collectively than they might otherwise be able to do individually. Groups that practice healthy interactive dynamics can influence others within the organization by their behaviors. Organizational behavior can be influenced from the bottom up as these healthier dynamics within the self-organizing groups are experienced and seen to effect positive results within the organization.

The positive effects of open facilitative communication skills are not limited to organizations in which we work. These are life skills that enhance our relationships wherever we may find them in our lives.

The Individual

How do we develop our individual views of reality that govern our behavior in life? Why do these various views of reality conflict with others' views of reality? In order to better understand others, we need to better understand ourselves first.

On a conscious level we have an idea of our ideal self: who we are and how we should ideally behave. When asked how we might respond to a hypothetical situation, this is the self we refer to. But underlying our ideal self that we present to the world in words is our true self, which reflects our core values through our behavior. What we say is not always what we do. Often we are not even aware that these two selves are not one and the same. Our core values are developed and modified throughout our lifetime. Behaviors we learn during our childhood from interactions with family and friends can unconsciously permeate our adult relationships. They've become so ingrained that we don't really think about them or recognize them. These behaviors are often triggered by the behavior of others in a reflex-like manner: unconscious and unaware.

Research points to our subconscious subtly directing our conscious responses (Douglas 2007). Our emotions are also critical in our ability to make decisions (Frith 2008; Khamsi 2007). When individuals are interacting with others, those interactions can trigger an emotional response that activates behavior patterns learned long ago in our childhood and in our youth. Often we do not think about how we are responding, we just react. This, in turn, can have a spiraling effect in the interaction with the other person. If the other person is not cognitively aware of their own emotional responses and interactive patterns, they, too, may respond reactively, thus setting up possible dysfunctional dynamics. Understanding that humans are creatures of habit behaviorally can help you consciously change the way you interact with others so that those interactions are more open, healthy, and productive.

Changing Our Behaviors

So how does one become more aware of these almost unconscious behavior patterns? How do we become more aware of the inconsistencies between our ideal and true selves? One traditional method that is very effective is keeping a behavioral journal. Much like a Dieter's Journal where you log everything you eat during the course of a day, with a Behavioral Journal you log your behavior throughout the day. You note the context of your behavior, the statements of others with whom you are interacting, and write what your actual comments and behaviors were (O'Neill 2007; Bacon and Spear 2003). This is a most effective tool for staying consciously aware of one's actual behaviors. Being aware is the key to changing one's behavior. With this journal you can begin to see patterns not only in your responses but in situations that trigger your responses. It also gives you a chance to explore other possible perspectives held by the other person.

If you assume the other person is acting out of respect and compassion for you, you are less likely to engage in unconscious negative communication patterns. Mistrust of others is often enhanced by assuming ill intent on the part of the other person. One can break that cycle if we assume good intent. More often than not, how we behave often triggers negative behaviors from others in a dysfunctional feedback loop. This does not mean that sometimes others do not have ill intent, but this open positive method of communication can create positive behavioral feedback loops that minimize misunderstandings in our interactions with others. This builds trust in our relationships.

Keeping the journal of your daily behaviors helps you to see where your behavior might actually create negative reactions from others. Being cognitively aware of our behaviors allows us the opportunity to slow down, think about what we're saying, and practice open positive communication skills in future interactions.

Views of Reality

Again, since we are inside a complex system and not external observers, we are limited in the reality that we understand. One of the greatest sources of conflict between individuals and groups is the impression that our view of reality is the only one—the only right one. Because of all of our limited positions inside the system, we think all of our views of reality are correct, when in fact, from a larger perspective most of our views of reality are short-sighted, at best, and probably wrong. Each individual view is but a slice of the whole reality. Putting this infinite number of views of reality together (an impossible task) would reveal a true reality. It is hard to step outside ourselves to understand that other people have had experiences in their life radically different and foreign to our own. It is hard to imagine or understand situations, experiences, and ideas that we ourselves have never experienced. And the experiences that we have in life affect the way we think and the way we behave.

The need to control others in order to get one's own way can have very negative consequences for interactions with others. Schwarz and Davidson (2005, 37) point out four main dangerous core values that can interact with rigid views of reality and negatively affect our behavior: (1) deciding on a goal by oneself and trying to control others to achieve that goal; (2) defining winning as getting what you want and treating any changes in your associated plans as losing; (3) trying to exert control by resisting the expression of contrary feelings by others; and (4) denying the underlying emotions and feelings associated with issues especially within yourself, and thus assuming you are acting in a very rational and logical manner.

Schwarz and Davidson (2005, 37) also identify a set of assumptions that combine with the core values listed above in reflecting a rigid view of reality: (1) your way of looking at a situation is the only way that reflects the one true reality; (2) your view of reality is the only right one and those who disagree with you are wrong; (3) since you have no underlying emotions regarding an issue, your motives are in the best interest of the group, and those who disagree with you have questionable motives; and (4) because people don't understand your perspective (the one true reality) your feelings are justified, instead of considering that your thinking influences your feelings and that your thinking may not encompass all possible perspectives.

As a result of these core values and assumptions, Schwarz and Davidson feel that people develop the following strategies to guide their actions in attempting to control others: (1) tell others what they should do; (2) keep your reasoning for your position concealed as you try to covertly lead others in the conversation to the conclusion you want them to draw; (3) don't ask about others' reasons for their positions for it may cause you to question your own reasons; (4) indirectly state your position in a question form that is designed to get others to see things your way (in the courtroom this would be called *leading the witness*); and (5) you tell yourself you are doing this to help yourself and others. These strategies are designed to try to take control of the situation.

By not revealing your reasons, you do not subject your viewpoint to cross examination by others, which might reveal flaws that you might find difficult to accept if you have a rigid view of reality. You don't ask others about their reasons in order to prevent them from being embarrassed when flaws and inconsistencies are revealed to the group because you believe any view that doesn't agree with yours has to be flawed. In addition, you don't want others' reasoning revealed because it might allow negative feelings from yourself or other members of the group to surface. Conflict is not considered to have any positive attributes. You also don't want others reasoning to be expressed because it might encourage others to question yours. All of this leads to a great potential for misunderstandings. It assumes everyone understands what everyone's talking about. Only through questioning why we feel the way we do and why we say the things we say are we able to clarify and understand what someone means (Schwarz and Davidson 2005, 38).

These attempts to control the group dynamics often result in unintended consequences. Instead of creating group unity, it often creates mistrust. Feelings and positions are suppressed, and defensiveness can often be increased. Learning, one of the key points of Richard Knowles's Process Enneagram© (2002) that enables a self-organizing system to be flexible and adaptable in an ever-changing environment, is greatly reduced because the feedback loops are essentially terminated. The opportunities to question and learn from each other are greatly reduced. Being unable to express one's feelings can increase stress and conflict. This in turn diminishes the motivation of individuals within groups and diminishes quality of life (Schwarz and Davidson 2005).

The ability to recognize that there are multiple valid views of reality or perspective enables one to more readily embrace uncertainty. Embracing uncertainty means giving up control. And this is one of the keys to understanding complex systems, as well as to being more flexible and adaptable in complex environments (Malone 2007).

Again, complex systems, particularly living human systems, cannot be controlled. Schwarz and Davidson point out that often when people realize that they are trying to control a situation, they will respond by doing the opposite, giving up all control. This response is not the most effective either. Schwarz and Davidson feel that this is exchanging one form of control for another. The core values of this model include: (1) everyone participates in defining the purpose, (2) everyone wins and no one loses, (3) expression of feelings, and (4) [suppression of] intellectual reasoning (Schwarz and Davidson 2005, 40; Argyris, Putnam, and Smith 1985). Here, the assumption is that the only way for people to learn is to come up with a correct answer themselves, but the catch is, only your perspective is the correct answer. If others in the group do not hold your perspective, then you have to guide them to your correct perspective. Here, again, control is being exerted. This model that supposedly gives up control obtains the same results as the control model: "increase misunderstanding, unproductive conflict and defensiveness, as well as reduced learning, effectiveness, and quality of work life" (Schwarz and Davidson 2005, 41).

Thus, this unilateral control model (summarized in Figure 5.9) is *not* the way to go.

Unilateral Control Model

Core Values and	Assumptions	Strategies	Consequences
• Achieve my goal through unilateral control	• I understand the situation; those who see it differently do not	• Advocate my position	• Misunderstanding, unproductive conflict and defensiveness
• Win, don't lose		• Keep my reasoning private	• Mistrust
• Minimize expressing negative feelings	• I am right; those who disagree are wrong	• Don't ask others about their reasoning	• Self-fulfilling self-scaling processes
• Act Rational	• I have pure motives; those who disagree have questionable motives	• Ease in	• Limited learning
		• Save face	• Reduced effectiveness
	• My feelings are justified		• Reduced quality of work life

Figure 5.9   Unilateral control model. (From Chris Argyris and Donald Schön. Theory *in* Practice: Increasing Professional Effectiveness. San Francisco: Jossey-Bass, 1974; Roger M. Schwarz and Anne Davidson; Roger M. Schwarz (ed). *The Skilled Facilitator Fieldbook*. San Francisco, CA: Jossey-Bass, 2005, 36.)

Instead, Schwarz and Davidson advocate what they refer to as the mutual learning model of Figure 5.10 as the only effective model for healthy group dynamics (Schwarz and Davidson 2005, 41). Similar to Knowles's Process Enneagram©, it reflects group counseling techniques to facilitate open and honest communication, which results in a more effective group interaction. The core values here include: (1) sharing all relevant information, including feelings, reasoning, and assumptions; (2) making free and informed choices without being manipulated; (3) obtain consensus and commitment from individuals in the group because the members were able to express their feelings and opinions and be heard; and (4) being nonjudgmental with others in the group through demonstrations of respect and compassion.

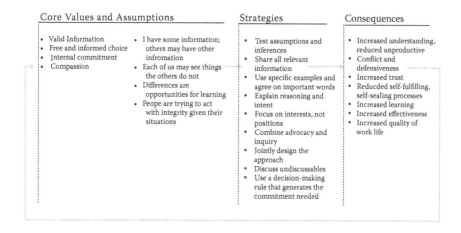

Core Values and Assumptions		Strategies	Consequences
• Valid Information • Free and informed choice • Internal commitment • Compassion	• I have some information; others may have other infromation • Each of us may see things the others do not • Differences are opportunities for learning • Peope are trying to act with integrity given their situations	• Test assumptions and inferences • Share all relevant information • Use specific examples and agree on important words • Explain reasoning and intent • Focus on interests, not positions • Combine advocacy and inquiry • Jointly design the approach • Discuss undiscussables • Use a decision-making rule that generates the commitment needed	• Increased understanding, reduced unproductive • Conflict and defensiveness • Increased trust • Reducded self-fulfilling, self-sealing processes • Increased learning • Increased effectiveness • Increased quality of work life

Figure 5.10   Mutual Learning Model. (From Roger M. Schwarz; Roger M. Schwarz et al. (ed). *The Skilled Facilitator Fieldbook*. San Francisco, CA: Jossey-Bass, 2005, 42.)

Respect and compassion comes from being able to embrace the uncertainty that there is more than one perspective regarding the reality we personally see. There is not just one right way. Schwarz and Davidson point out: (1) each of us have a different perspective on the situation and may have information of which others are not aware; (2) we cannot truly know how others see the world and their reality; (3) diversity of perspectives is an opportunity to learn from others and expand our own views of reality; and (4) most people are behaving with honest motivations within their personal views of reality (Schwarz and Davidson 2005, 43, 44).

The results of the mutual learning model are similar to those achieved by Knowles's Process Enneagram©: increased understanding, communication, trust, effectiveness, learning, and meaning in life and motivation.

Both of these models tap into deep understandings of individual and group dynamics, which enable them to be very effective techniques.

Human Systems:—What Makes Them Complex

People communicate and interact with one another, which creates a complex system. Any action that one person takes can impact others in multiple ways: they can be seen or unseen, have unanticipated consequences, and the effects may not be known immediately. Peter Senge (1990, 57–67) points out several Laws of Systems that also reveal fundamental concepts of human nature that are constant and don't change. The ones that overlap include the following:

Today's Problems Come from Yesterday's Solutions

Senge points out that some solutions merely hide the problem by shifting it to another area of a complex system. This parallels what we know about human behavior, too. People will often camouflage or deny what the real issues are that they are facing in their life. Sometimes it's too painful to acknowledge and is hoped will go away on its own if we don't deal with it. Or they will try to forcibly control the negative symptom of the problem in their life. In counseling psychology, this dynamic is often demonstrated using a balloon metaphor. Here a person's life is represented by the balloon. If we are experiencing difficulties in our lives, often we will clamp down on the bulge or problem that's protruding from the balloon, only to find out to our dismay that another bulge pops out elsewhere on the balloon. Similar to what Senge is saying about shifting problems to other areas of a complex environment, the balloon metaphor demonstrates that the interaction among people results in uncontrollable and unpredictable consequences. Similar to shifting the problem you're trying to control to other areas of complex system, not understanding the complex web of interactions of people, and the ripple effect from those interactions, can result in unexpected problems developing elsewhere.

The Harder You Push, the Harder the System Pushes Back

Systems, including human systems, are resistant to change. Often, when you try to force people to change or do what you want them to do, you are met with resistance that requires greater effort to overcome. On an individual level we see the same difficulty. Trying to make changes in our own personal behavior requires diligent constant effort or else we fall back into old familiar paths and do not change. It is as if our own physiology is pushing back against us. This same dynamic is seen on a larger scale in group behavior. Also, similar to behavior found in children, when groups resist change, they often act out in a more intense fashion. Many times the problems or behaviors become much worse, eliciting greater control methodologies to be employed.

Behavior Grows Better before It Gets Worse

Here, Senge is referring to treating the symptoms of a problem and not really curing the issue that causes the problem. This is often seen in counseling situations, too, where the underlying hidden interconnected dynamics are at work. Dealing with problems as if they were simpler linear ones often results in treating just the symptoms and not truly dealing with a complex dynamics. In the short run it looks as if things are getting better, but in the long run the problem will persist and often get worse.

The Easy Way Out Usually Leads Back In

This refers to people's reluctance to change. We have established patterns of behavior that are comfortable and almost instinctive for us. It is always easier to unthinkingly follow a well-worn path even if that path enables a problem to persist.

Faster Is Slower

Senge says complex systems are unique and have varying rates of change. Due to the complex nature of the interconnections, it takes time to understand which solution might work best over the long run. Similar to dealing with human dynamics, quick fixes seldom work well for long. It takes time to understand how the interconnected dynamics work and how interventions at different levels of these interactions might impact the system as a whole.

Cause and Effect Are Not Closely Related in Time and Space

On the surface, our world in many ways seems linear. If you rap on a door with your knuckles you instantly hear the resulting knock. The action is related closely in time and space. But in complex human systems, where much of the interconnected dynamics is unseen, the consequences of one person's action or a change within a group might not be seen for years.

Small Changes Can Produce Big Results, but the Areas of Highest Leverage Are Often the Least Obvious

Similar to the metaphor of the butterfly in Central America flapping its wings and subsequently setting off a hurricane in the Atlantic Ocean, so too can small changes in complex human systems that are interconnected cause a large impact, especially when these actions are taken at critical connecting points (individuals) within the system. The difficulty here is that many of these interconnections, and consequently these critical points, are not readily observable.

Dividing an Elephant in Half Does Not Produce Two Elephants

Complex systems cannot be deconstructed. Unlike taking a vacuum cleaner apart and examining the individual parts to try to determine what's causing a malfunction, the nature of a complex system depends on the interaction of all the parts together. This is true, too, in understanding the problems individuals face in their daily lives. It is the interactions that cause the issues that need to be addressed. Complex human systems are co-created, involving the interaction of multiple perspectives or views of reality.

There Is No Blame

It is common for people to blame others or circumstances when things go wrong, but systems thinking says we are all inside the system and that there is no outside observer position. From a counseling psychology perspective that idea reinforces that the individual is responsible for their own actions. The solutions for problems that emerge from within a complex system can only be found within that system itself, i.e., within the interactions found within that complex system and the systems interactions with its external environment. Individuals may not have any control over the behavior of others within the complex system, but they do have control over their own behavior.

Summary

All of the points referred to in the previous section reinforce the idea of interrelatedness and feedback loops within systems, especially human systems. Understanding this interrelatedness requires a slowing down in our decision-making processes and allowing time to look at the complexity of these human systems. Trying to find quick solutions to problems often negates engaging concerned individuals in the decision-making process.

If people are unable to express their concerns and opinions about issues the group is facing, they will resist changes that may be imposed upon them. This resistance can take many forms, one of which is poorer communication among members of the group. Often you can achieve greater likelihood of consensus and reduce the negative impact of conflict if people feel their voices have been heard.

Quick solutions may seem to work in the short term, but resistance and resentment can grow under the surface, resulting in problems down the road. Often a change in the way we think about a problem or an issue can be the small change that has a large impact on a complex system. From a counseling psychology perspective, changing the way we think can change our behavior, and changing our behavior can lead to a dramatic impact within an interconnected complex system.

Sometimes, too, just having a different perspective on a problem resolves the problem or makes more options become available. Just as Senge's Laws of Systems are interrelated, so too are the dynamics found in complex human systems. In order to help groups self-organize and solve the issues they face, one needs to think in terms of all the hidden and visible dynamics that are interacting within the whole system.

It is important to understand how the interactions of the system operate as the group establishes the ground rules by which they will work together, as well as when you use your facilitative group skills in your conversations with others in the group. We do not operate in isolation and what we say, as well is what we do, can be interpreted in a multitude of ways through the filters of others' views of reality. The consequences of those interactions may not be readily visible or immediately felt.

Using co-created ground rules for how individuals in the group interact can minimize some of the destructive qualities of conflict. But for the rules to be effective and the conversation to flow freely, healthily, and productively, individuals need to understand how the complex social interactions impact the group and their collaborative goals. They need to also understand that our thoughts can change our behaviors just as our behaviors can influence our thoughts (Mithaug 1991, 84). Learning is the key to being able to change, as well as the key to adaptability. And a willingness to change depends on how much we understand and accept.

Chapter 6

Wicked Problems and MUVEs: Understanding Human Interactions through Multiuser Virtual Environments

B. G. McCarter

Welcome to a dramatically evolving world! We have seen an unprecedented explosion of information the last several years, bringing us to 2012. (For example, refer to Appendix E.)

Our efforts to develop technology to greater aid our succeeding in this ever-changing landscape are resulting in the Information Age giving way to the Conversation Age. People are interacting on a global scale, often in synchronous time. They are sharing information, self-organizing to move goals and organizational missions forward, and emergence (unpredictable and unplanned consequences of those interactions) is driving even more complexity.

Today, it is about building relationships and trust, through sharing information and enabling flexible collaborations on a global scale. This is redefining who we are and how we work together. And how we interact is shaping our perspectives and attitudes.

Dynamics of Living Human Systems at Work

In today's world it is important to understand these dynamics and how they enable self-organization (a key skill in enabling organizations and groups to be flexible and agile). The world is hyper-connected and the conversation has begun. New processes are needed to enable organizations to move forward successfully in this new structure.

As discussed earlier in this book when we discussed the Process Enneagram (refer to Chapter 4, in particular), there are three basic dynamics involved in the living human systems: identity, information, and relationships (refer to Figure 6.1), and their importance is being keenly felt in today's complex changing world.

Identity involves

■ one's sense of self,
■ one's unique perspectives,

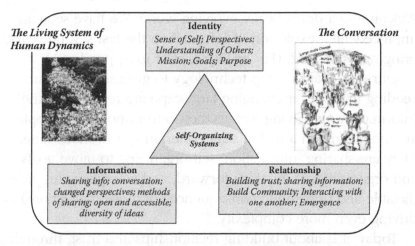

Figure 6.1   Human dynamics: the living system.

- our understanding of others, and
- our mission, goals, and purpose in life.

Information refers to

- the act of sharing information,
- engaging in conversations with others,
- adjusting our perspectives as a result of new information,
- developing methods of sharing information,
- being open and accessible to new information, and
- the diversity of ideas that occurs when you are sharing information in a hyper-connected environment (which enables complexity and facilitates creativity).

Relationships encompass

- the act of building trust through the sharing of information,
- building community through shared networks,
- interacting with one another, and
- the unpredictable emergence from human interactions.

Due to global financial considerations and a change in the way people process information, people and organizations have to find new ways to meet, collaborate, and learn. In that search, virtual worlds are proving to be an effective medium. Virtual worlds (also known as *multiuser virtual environments* [MUVE] and *3D learning environments*) mirror the complex human dynamics of our physical world. The psychology behind virtual worlds is one of the reasons virtual worlds are so effective.

Psychology of Virtual Worlds

Not only are virtual worlds a great place for learning, for collaboration, for research and development, for prototyping, and data analysis, but they are also able to impact us psychologically, behaviorally, and physically.

It is a medium where we truly extend ourselves (our identity or sense of self) and meld into the environment. As a result, organizations that use this environment need to begin to pay special attention to the psychology of design involved with the avatars used, as well as the interactive space of the virtual world.

The psychology of the avatar and our interactions with others in the space is what creates presence, that all important and elusive element from a sense of a shared space and time with others that builds trust and cooperation, enabling collaborative efforts that transcend time and place in our hyperconnected world.

This is an area that has not received focus in past research, but one that is gaining importance and attention.

One paper on this issue was recently published in the *Journal of Virtual Worlds Research* titled "Who Am I—and If So, Where? A Study on Personality in Virtual Realities," by Benjamin Gregor Aas, Katharina Meyerbröker, and Paul M. G. Emmelkamp (http://journals.tdl.org/jvwr/article/download/777/707). This paper examined the stability of personality traits in virtual worlds and found that personality traits remained stable as users entered virtual worlds.

However, another study seems to suggest that attention needs to be paid to how we design our avatars, as their results seem to suggest that how we represent ourselves in virtual worlds affects our behavior in our physical world. Refer to "The Proteus Effect: The Effect of Transformed Self-Representation on Behavior" (http://citeseerx.ist.psu.edu/viewdoc/download?doi=10.1.1.134.6224&rep=rep1&type=pdf).

We become or behave like that which we put on, much as our behavior may change depending on the type of people with whom we associate. If we are affected by the behavior and lifestyles of friends of friends of friends (three degrees of separation http://tinyurl.com/7fpamh), how much more easily affected are we by stepping into an avatar that is an extension of ourselves with inherent feedback loop capability?

Another study published in the *Journal of Virtual Worlds Research*, "The Effects of Avatar Appearance in Virtual Worlds" (http://journals.tdl.org/jvwr/article/download/843/706), supports this idea and points out that one's avatar's appearance does indeed affect our behavior.

This is further evidence of the mind seamlessly embracing virtual worlds and that the possible feedback loop represents the success and increased applications of psychological therapies in virtual worlds. Psychotherapy and various physical therapy treatments dealing with pain and burn patients would not be so successful in a 3D immersive environment if the mind did not reach out and immerse itself in the environment.

In fact, the video *Snow Worlds* (http://youtube/jNIqyyypojg) talks about how immersing the burn victim in the virtual world during therapy and bandage change is able to reduce the pain because the virtual environment is able to physically dampen the pain centers of the brain. In addition, Club One Island has done research to prove that engaging in a weight reduction and physical fitness program in a virtual environment actually leads to behavioral changes that result in weight loss. (Refer to "Weight Loss Success in a 3-D Virtual World," http://www.sciencedaily.com/releases/2011/06/110603102736.htm)

Virtual worlds or 3D immersive environments comprise a powerful medium as opposed to a brick-and-mortar space that does not touch who we are. Virtual worlds are nebulous spaces that allow our minds to extend themselves as never before.

The mind wants to reach out into its environment as is known by anyone who has ever used a stick to explore a dark hole and can feel what the end of the stick touches. Virtual environments and the avatars that inhabit them are extensions of ourselves, and we need to be mindful of this as we design these incredibly powerful spaces that are becoming more commonly used in educational, organizational, and personal lives.

Every day more research in this topic is being published to help give us a better understanding of the deeper impact of virtual worlds (refer to Appendix F).

Next Level of Interaction and Learning

Virtual worlds are part of the next level in the transformation of how we learn and collaborate in our increasingly complex world.

Because today's hyper-connected complex world is challenging the way we collaborate and work together, it is necessitating a new way to learn, as well as relearning how to interact positively with one another.

The traditional Industrial Revolution–style classroom learning is giving way to minds that learn best through lateral thinking (connecting the dots), and kinetic hands-on learning; and virtual worlds are providing a powerful platform through which learning in today's ever-evolving complex world is finding greater applications.

- Participants are able to experience abstract ideas that can't be readily experienced in our physical world.
- They are able to kinetically learn skills and practice them in an environment that enables complex emergence (those unpredictable experiences that occur when people get together and interact).
- They can learn specific skills or information at a deeper level and then practice those skills in dynamic changing environments where *how* they use those skills is not laid out by the instructor. The participant learns to use their knowledge and skills in changing contexts.

And these classrooms are increasingly utilizing collaborative learning. Learners in virtual environments are learning not only the skills needed to succeed today, but they are increasingly being challenged to learn those skills in unscripted multiuser virtual environments, working with others to achieve goals and solve problems.

An understanding of the inherent dynamics of the living human system is needed to address "wicked problems" that arise in multiuser virtual environments: the impact of human interactions and the unpredictable complex systems they enable. Virtual worlds do indeed mirror the physical world.

As a result, learners today need to learn skills they can utilize in ever-changing contexts that they have never seen before or even imagined, and be able to utilize those skills in collaborative environments.

Today, learning environments need to incorporate an understanding of

- group dynamics,
- complexity (those unseen dynamics that have dramatic impact when you least expect it!), and
- collaboration skills.

There are four basic elements that have a significant impact on virtual world learning designs with wicked problems in mind. They enable a deeper learning that the participant truly owns and that helps to achieve a greater likelihood of collaborative success.

These four interrelated basic elements include

- group dynamics and the impact of those interactions,
- our sense of identity (and the avatar is an integral part of this in virtual worlds),
- the relationships we develop, and
- the power of storytelling and narrative structures to help us understand complex situations (refer to Figure 6.2).

These elements continue to keep in mind the three elements of the living human system: identity, information, and relationships (refer to Figure 6.3).

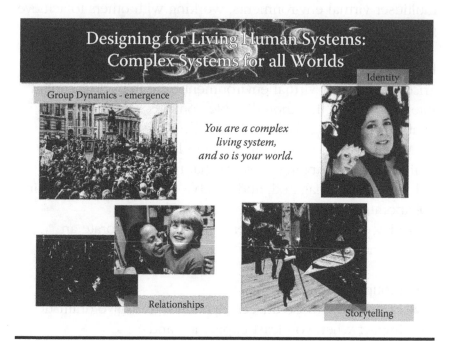

Figure 6.2   Designing for living human systems.

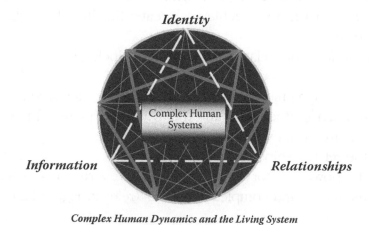

Figure 6.3   The living human system.

Narrative Structures and the Underlying Psychological Dynamics

A renewed appreciation of the power of narrative structures is arising as it is being incorporated into virtual environments. A greater understanding of abstract ideas or concepts is enabled as one is able to become a kinetic part of a story. Historically, storytelling (Herodotus, Shakespeare, and modern theatre and cinema, for example) helps us to make sense of our world, especially the complex dynamics we experience.

Narrative structures impact the three core elements in living human systems in the following ways:

- **Identity**
    - Allows us to try on the roles of others and experience their perspectives
    - Helps us see issues from different perspectives, and make decisions in a different environment
    - Influences our sense of self
    - Shapes our emotions and our actions
    - Enhances or changes our perceptions
    - Involves our feelings and emotions
- **Relationships**
    - Develop rules of behavior
    - Develop community through a shared story
    - Influence our group identifications
    - Impact our interactions with others
- **Information**
    - Allows us to see consequences of our actions
    - Help us to understand complex dynamics
- Changes the way we think

These elements are the heart of the Process Enneagram, which helps groups and individuals work together to solve problems in ever-changing complex environments. Deeper

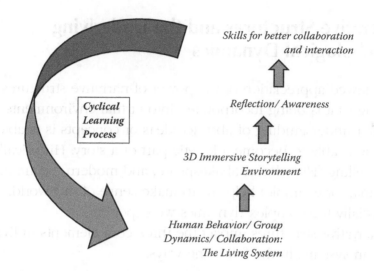

*Skills for better collaboration
and interaction*

*Cyclical
Learning
Process*

Reflection/ Awareness

*3D Immersive Storytelling
Environment*

*Human Behavior/ Group
Dynamics/ Collaboration:
The Living System*

Figure 6.4   Cyclical learning process.

understandings of these variables make collaborations easier
and more likely to succeed. By incorporating these four basic
elements into the virtual world's design for immersive learning,
one is able to facilitate a cyclical learning process. One learns
positive collaboration and human interaction skills, continues
to experience group dynamics, steps into a narrative structure
or story to gain a greater understanding of diverse perspectives; reflection and awareness continues to build, and better
collaboration and interaction skills are learned. The cycle continues in an iterative process, just as the Process Enneagram
does (see Figure 6.4).

The Information Age is giving way to the Conversation Age,
and the power of immersive 3D multiuser environments are
helping us work together and achieve our goals despite our
differences in personality, childhood history, culture, geo-
graphic location, or even generation.

Appendix A

Mini-Lexicon of Selected Terms

Informal definitions of specific terms used in this chapter are provided here for reference. The definitions, features, and notes are taken from a larger lexicon published by Taylor & Francis in the form of a "Wikipedia, http://enterprise-systems-engineering.com/phpwiki/" For ease of navigating the "soft-copy" versions of this lexicon, there is a link associated with each bolded term within a definition. The soft copy has links to other sources in some instances for those who want to learn more about certain terms.

Complexity: A technical term qualitatively describing the ultimate richness of an entity that (1) continuously evolves dynamically by organizing its own internal relationships; (2) requires **multi-view** analysis to perceive different non-repeating patterns of its behavior; and (3) defies methods of pre-specification, prediction, and control. (See **variation** and **selection**.)

Features

Complex entities possess attributes that cause them to evolve naturally without outside intervention. It is also not possible to pre-specify or predict completely and accurately what will happen with complex entities, even when one intervenes from the outside with a specific purpose. The attribute of complexity is usually associated with the property of instability. Furthermore, it is not possible to replicate complexity exactly. Each complex instance is unique. Increasing a system's complexity implies its potential behavior will display more variety, nuance, and depth. A system can become so complex that its state approaches that of chaos—and may even transition into chaos—making its nature even more difficult to understand. A system might also evolve with diminishing complexity, trending toward stability. This trend may continue to the point that the system might better be described as deficient in variety and richness, uninteresting, possibly stagnant, or even boring. The challenge in either case is to attempt to shape the environment of a complex system by continually introducing variety and selection [akin to Ashby's (1956) "Law of Requisite Variety"]. This enables a system to become even more complex, yet avoid chaos or stagnation.

Notes

Many people use the term *complex* as a synonym for anything that is complicated and difficult for a typical human being to understand. Although this is often appropriate in the English vernacular, when used in the context of enterprise systems engineering (ESE), *complexity* implies discerning in much greater depth the matter being considered.

Complex system: An **open system** with continually cooperating and competing elements. (See **enterprise**.)

Features

This type of system continually evolves and changes its behavior, often in unexpected ways, according to its own condition and external environment. Changes between states of order and chaotic flux are possible. Relationships among its elements are imperfectly known and are difficult to describe, understand, predict, manage, control, design, and/or change.

Notes

This suggests examining the role of a system's boundary in differentiating between open and closed. A closed system is merely a system that has been defined with respect to a boundary that contains the totality of its interactions. Inside that boundary, that same system might look open. On the other hand, even an open system has a boundary; otherwise, there would be no "external" to define or identify it. Defining the bounds of a system is a critical first step in any systems engineering process. It is sometimes possible to make the system open or closed by appropriately defining the boundary. A complex system is not merely complicated. It is nonlinear. And chaotic behavior can be an intrinsic property of the system that connotes the sensitivity of the system to perturbations of the initial conditions. When a system is then also given inputs from an aggregation of random processes, the result appears complex; a system that is predominately linear *can* be predictable, even if the inputs are random. A complex system is *not necessarily* an enterprise.

Engineering: Methodically conceiving and implementing viable solutions to existing problems.

Note

This definition is not meant to imply that the problems are always solved.

Enterprise: A **complex system** in a shared human endeavor that can exhibit relatively stable equilibriums or behaviors (homeostasis) among many interdependent component **systems**.

Features

An enterprise may be embedded in a more inclusive complex system. External dependencies may impose environmental, political, legal, operational, economic, legacy, technical, and other constraints.

Notes

An enterprise usually includes an agreed-to or defined scope or mission and/or a set of goals or objectives. Note also that this definition is meant to be limited. *Not included* here is a recipe for a successful enterprise. That is a different topic: enterprise engineering and enterprise systems engineering.

Enterprise systems engineering (ESE): A regimen for **engineering** successful **enterprises**. For more information, refer to the papers "A Framework for Enterprise Systems Engineering Processes, (http://www.mitre.org/work/tech_papers_06/06_1163/index.html)" "Engineering Enterprise Systems: Challenges and Prospects, (http://www.mitre.org/work/tech_papaers_06/06_0324/index.html)" "Practicing Enterprise Systems Engineering," and "On the Pursuit of Enterprise Systems Engineering Ideas. (http://www.mitre.org/work/tech_papaers_06/06_0756/index.html)"

Features

ESE is systems engineering that emphasizes a body of knowledge, tenets, principles, and precepts concerning the analysis, design, implementation, operation, performance, and so forth of an enterprise. Rather than focusing on parts of the enterprise, the enterprise systems engineer concentrates on the enterprise as a whole and how its design, as applied, interacts with its environment. Thus, ESE avoids some potentially detrimental aspects of traditional systems engineering (TSE), such as concentrating on parts of the system and their behavior in isolation. In contrast, an ESE approach focuses on how those parts interact within the system and with the outside environment.

Notes

Here *regimen* means a prescribed course of engineering for the promotion of enterprise success. Although most people would not bother to engineer anything without attempting to make a success of the effort, some techniques applied to enterprises, such as reductionism, can be unsuccessful. (See **work breakdown structure**.)

Environment: What embeds and surrounds any **system**.
Granularity: The ability of a person to discern and discriminate individual items of a conceptualization. (See **view**.)

Notes

Granularity is akin to a capability to observe details; it is like resolution. Subsets of detailed items will likely include arrangements or patterns, some of which may not be discernible in other views.

Mindset: What currently captures an individual's attention in a conceptualization. (See **view**.)

Note

Mindset is akin to one's cognitive focus, which may observe or contemplate (e.g., within one's **scope** and with the associated

granularity) a single object, pattern, notion, or idea, or a collection of such elements.

Procedure: An acknowledged or intentional way of doing things to achieve a desired goal. (See **process**.)

Note

A procedure can be considered a template for behavior or a specific instantiation or tailoring of a more general process.

Process: A relatively generic description (compared to a **procedure**, at least) of how one does things to accomplish a desired outcome or set of goals.

Scope: What is included in an individual's conceptualization. (See **View**.)

Notes

Conceptualization is akin to perception (i.e., visualization). Specific analogies to scope are the field of view of a camera or, more appropriately here, the "mind's eye." When one sets or determines scope, by definition, everything else not in scope is "abstracted out" (i.e., not "seen" by that individual, at least in that view), because those things are not relevant to the person's intended present state of being (i.e., purpose).

Selection: In **enterprise systems engineering**, selection is the act of restricting or limiting choices in the **environment** to shape a solution set. (See **Variation** and **Complexity**.)

Note: Too much selection can lead to stagnation, which is, in most cases, not desirable.

System: An interacting mix of elements forming an intended whole that is greater than the sum of its parts.

Features

These elements may include people, cultures, organizations, policies, services, techniques, technologies, information/data, facilities, products, procedures, processes, and other human-made (or natural) entities. The whole is sufficiently cohesive to have an identity distinct from its environment.

Note

In general, a system does not necessarily have to be fully understood, have a defined goal or objective, or have to be designed or orchestrated to perform an activity. However, in the present definition, *intended* means an understood or defined goal or objective and designed or orchestrated to perform a useful activity.

Systems engineering: An iterative and interdisciplinary management and development process that defines and transforms requirements into an operational **system**. For more information, see the INCOSE website (www.incose.org).

Features

Typically, this process involves environmental, economic, political, and social aspects. Activities include conceiving, researching, architecting, utilizing, designing, developing, fabricating, producing, integrating, testing, deploying, operating, sustaining, and retiring system elements.

Notes

The customer for or user of the system usually states the initial version of the requirements. The systems engineering process is used to help better define and refine these requirements. Further, the requirements often change as new decisions are made as a result of systems engineering. This definition does not imply that a successful system is always realized. The word "integrated" is not included in this definition because systems engineering efforts are not always that well integrated.

Timeframe: The time interval of an individual's conceptualization. (See **view**.)

Note

Timeframe is akin to the temporal component of one's conceptualization; in other words, the timescale over which it occurs.

Variation: In the context of **enterprise systems engineering**, the act of allowing or encouraging perturbations in the **environment** with the intention of creating a richer variety of potentially attractive solutions. (See **selection** and **complexity**.)

Note

Too much variation can lead to chaos, usually an undesirable state of affairs.

View: Any combination of **scope, granularity, mindset,** and **timeframe**.

Features

A change in any one of these elements will result in a change of view (i.e., what one can perceive or understand).

Notes

The limitations of the human brain make it practically impossible for a person to essentially grasp the underlying "reality" of any situation. Rather, each person viewing something focuses his or her mind on a mental snapshot or perspective of a situation. Each person understands it only to a certain extent (or scope), with its associated level of granularity (detail), abstracting out what appears to be irrelevant for one's own particular viewpoint. Even someone totally unfamiliar with ESE can identify with the saying, "If you can't change the situation, change your attitude." Attempting to take a fresh look at something familiar from unfamiliar points of view can be a useful device to gain further understanding of a system.

Work breakdown structure: A divide-and-conquer **procedure** to manage work effectively.

Notes

This procedure is quite entrenched in the TSE process and is consistent with reductionism and constructionism. However, work breakdown structures usually are not as effective in system of systems engineering, ESE, or CSE, because these systems tend to change continually, in spite of conventional engineering efforts.

Appendix B

INCOSE Working Group Sidebar on Complex Systems

The SSEG was reconstituted as the Complex Systems Working Group (CSWG) in February 2008 under the leadership of Ms. Sarah A. Sheard, an INCOSE fellow (sheardssheard@cox.net or sheard@3MilSys.Com).

The following are my answers to several questions raised by INCOSE Fellow Jack Ring:

1. *What distinguishes complex systems from other kinds of systems?* A complex system includes at least one autonomous (can act independently) agent that may respond to stimuli from other elements of the complex system or that system's environment.

2. *Is more complex always better?* This question suggests that complexity is one dimensional, with an attached measure that makes "more" or "less" meaningful. Indeed, one can think of a one-dimensional complexity scale, where the lowest level denotes "complicated" and the higher levels become increasingly more complex. More generally, complexity can be viewed as multi-dimensional. A complex system, then, might be depicted in some abstract n-dimensional space, where n is unbounded or even infinite, perhaps. Conceptually, one also can think of an n-dimensional complex system

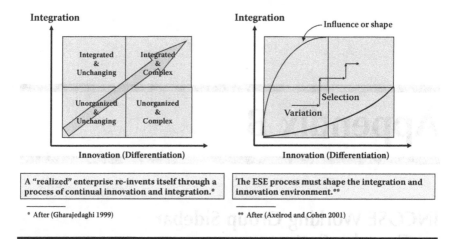

A "realized" enterprise re-invents itself through a process of continual innovation and integration.*

The ESE process must shape the integration and innovation environment.**

* After (Gharajedaghi 1999)

** After (Axelrod and Cohen 2001)

Figure B.1   Systems engineering process moving to enterprise systems engineering.

as being "projected" into fewer than *n* dimensions. A simple two-dimensional example, where innovation and integration are the two degrees of freedom, is depicted in Figure B.1. (Refer to the definitions of *variation* and *selection* in Appendix A.) Here one can envision complexity "increasing" as one moves from the lower left to the upper right. In this sense, more complexity (i.e., more innovation and more integration, simultaneously) is "better," assuming one is able to influence or shape outcomes in this space to avoid chaos (too much innovation) or stasis (too much integration).

3. *What distinguishes a complex system from the systems engineering of a complex system and from complex systems engineering of a complex system?* Systems engineering (SE) (refer to the definition in Appendix A) constitutes a purposeful set of activities or actions, by people, applied to a system; thus, SE is distinguishable from the system itself, whether it be complex or not. Complex SE is a subset of SE. Two different views of several engineering terms (refer to definitions in Appendix A) are depicted in Figures B.2 and B.3. Figure B.2 provides a set theory

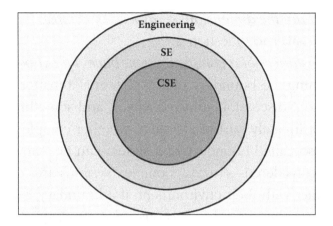

Figure B.2   Set theory view of engineering disciplines.

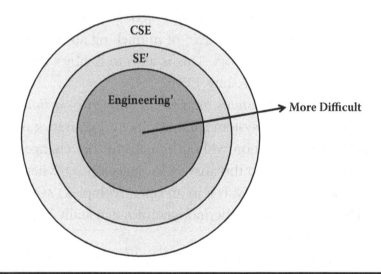

Figure B.3   Degree of difficulty view of engineering disciplines.

point of view, comparing complex SE (CSE) with SE and engineering in general; Figure B.3 provides a degree of difficulty viewpoint.

4. *Is SE, the act, one aspect of a complex system?* No. Refer to the answer to Question 3.

5. *What are the dimensions or aspects of complex?* Refer to the answer to Question 2.

6. *What are the dimensions or aspects of complexity?* Refer to the answer to Question 2.

7. *Do instances of a complex system have a fixed boundary?* Defining the boundary of any system is a matter of perception, discussion, and consensus, and something that is not innately part of a system, whether complex or not. The so-called boundary of a system can be—and is usually considered—"fuzzy." Complex systems are "open" and interact with their environment; if a boundary is defined, it should be viewed as "porous," with "energy" flowing across the boundary in all directions.

8. *Must the SE that models a complex system be more complex than the complex system it models?* Yes, if one wants such a model to be capable of mimicking all of the complex system's behaviors. This is akin to Ashby's Law of Requisite Variety (Ashby 1958).

9. *Do all complex systems increase in entropy over time?* Not necessarily. The system can atrophy and go into stasis [a "frozen" condition without (much) further change], for example. (Refer to the answer to Question 2.) What does entropy mean, anyways, in an open (complex) system? (The second law of thermodynamics is usually applied only to closed systems.)

10. *Do some complex systems self-simplify?* This depends on what one means by simplification. Whether something is simple or not depends on one's perception and viewpoint. Multi-view analysis is often required to approach a deeper understanding of a complex system. Sometimes a pattern (e.g., "strange attractor") will be seen that implies a breakthrough to simplicity at the other side of complexity, if you will. "Healthy" complex systems self-organize, often to attain a higher level of robustness, survivability, or whatever.

11. *Are there measures of effectiveness for the x in complex?* Conceivably, yes, but these measures are not necessarily inherent in the complex system but are imposed by observers. (Refer to the answer to Question 2.)

Duane Hybertson is another member of the SSEG (CSWG) and a fellow staff member of the MITRE Corporation. The following are his thoughts (with which we basically agree), as provided in an e-mail dated 13 June 2008:[*]

Traditional Systems View

Traditional systems engineering (TSE) has a reasonably well-understood stance toward the systems it engineers. The TSE view of system characteristics is that:

■ They are large-scale machines that have relatively predictable behavior.
■ They have components from multiple disciplines, such as power sources, electrical and mechanical systems, computer hardware and software, control systems, safety, security, and communications.
■ Their components are machines or mechanistic elements, which can include any physical material used to construct systems such as aircraft, vehicles, buildings, dams, and equipment, as well as traditional software and electronic items such as computers. This scope was clearly articulated in Goode and Machol (1957, 5), who acknowledged common points of interest with social, biological, and ecological systems, but excluded them from the scope of SE.

---

[*] This material is used with permission. Many of these ideas were later published (Hybertson 2009).

- Their components exclude people. This is implied by the previous item, but is important enough to point out separately. The claim is sometimes made in TSE that systems include people and processes. However, in TSE, people usually play one of three roles, all of which reside *outside* the system and are treated as distinct from the system. The first role is systems engineer; the second is project manager, i.e., manager of a SE project that engineers a system; the third is user or operator of a system (addressed in the areas of human–machine interaction, human factors, and ergonomics). Therefore, in the traditional view, people are not system components.
- They are relatively stable. They change only if the owner decides to change them. Left to their own devices, they remain static or they deteriorate through normal use.
- Development and operation are separate phases of the system's life.
- They are designed and organized by an external designer—i.e., by the systems engineer.

The characteristics described above reflect a system-as-machine metaphor, a mechanistic view of systems.

COMPLEX SYSTEMS VIEW

Complex systems tend to exhibit one or more of the following characteristics:

- They are thought of as organisms that have relatively unpredictable behavior.
- They have a variety of components that are highly interconnected, and the behavior of the

whole system depends in a significant way on the interactions among components.

- The systems, and their components, are to some degree autonomous, acting on their own.
- Their components are considered to be people or organic elements, which can include anything in the realm of biology, psychology, and sociology. In addition to humans, examples include cells, organs, and social groups such as organizations and societies. Other examples, such as self-modifying learning software and cellular automata, are not necessarily organic but can be modeled as organic or autonomous.
- They are adaptive and learn, grow, and evolve in response to their environment. Left alone, they grow and improve. They continually change, but they also have significant aspects of stability. They also change their environment.
- Their development and operation are conflated into an ongoing evolution process in which change occurs while the system is in operation. In many cases, there is a formation period (such as the gestation period of animals), followed by a longer evolution period.
- They are self organizing, and the whole system emerges over time as a result of cooperation and competition among components.
- They have expanding scope: They go beyond technical issues to social, political, and organizational issues.

These characteristics reflect a system-as-organism metaphor, an organic view of systems.

Clearly these characteristics are in contrast to those described earlier from the perspective of TSE.

The contrast of the organism perspective with the machine perspective is reflected in certain changes in mindset, where constructs that are viewed negatively in TSE are embraced as positive or at least potentially positive in a complex systems approach. These constructs include:

- *Change*: Change is something to be avoided in TSE, or at least carefully controlled. In complex systems, change is a natural process. It can be positive, negative, or neutral, but in many cases, change reflects learning, growth, improvement, and adaptation
- *Risk*: TSE attempts to minimize or avoid risk. A complex systems approach examines opportunity, the flip side of risk, as a positive potential for significant gains.
- *Uncertainty and lack of control*: TSE tries to maintain maximum control and to minimize uncertainty. With complex systems, engineers can yield some of that control to systems that exhibit autonomy and self organization, which reduces the amount of necessary instructions or explicit specifications. Instead of the engineer having to anticipate everything for the system, let the system work things out for itself as it goes along.
- *Contradiction and paradox*: TSE seeks to achieve consistency and avoid contradiction, paradox, tension, and contrast. Complex systems accept contradiction, tensions, and contrasts as representing balance and diversity, in the spirit of yin and yang. Contrasting elements are reconciled in part with the use of views.

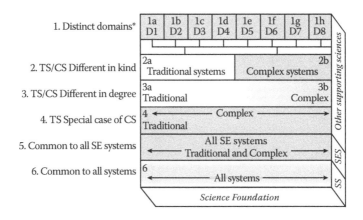

Figure B.4 Relationships between traditional systems and complex systems.

COMBINED VIEW

I suggest that many, if not most, systems engineered in the future SE will be hybrid systems that include both people and machines. The primary cause of this will not be that the systems will change drastically. Rather, the big change will be that SE *recognizes* systems for what they are. We have had a significant blind spot, trying to treat our systems as mechanistic, and the future embrace of a broader science foundation for SE will remove the blind spot and enable us to use more appropriate models, including organic CSS models, for the actual systems we are attempting to engineer. Figure B.4 reflects the science foundation and the different relations between traditional systems (TS) and complex systems (CS).

Appendix C

Quotations from *The 48 Laws of Power*

Snippets from the book's preface prepare the stage:

> … being perfectly honest will inevitably hurt and insult a great many people, some of whom will choose to injure you in return. No one will see your honest statement as completely objective and free of some personal motivation. And they will be right ….

> If the world is like a giant scheming court and we are trapped inside it, there is no use in trying to opt out of the game. That will only render you powerless, and powerlessness will make you miserable. Instead of struggling against the inevitable, instead of arguing and whining and feeling guilty, it is far better to excel at power.

> … power's crucial foundation, is the ability to master your emotions.

> … what separates humans from animals is our ability to lie and deceive.

If deception is the most potent weapon in your arsenal, then patience in all things is your crucial shield. Patience will protect you from making moronic blunders. Like mastering your emotions, patience is a skill—it does not come naturally. But nothing about power is natural; power is more godlike than anything in the natural world. And patience is the supreme virtue of the gods, who have nothing but time. Everything good will happen—the grass will grow again, if you give it time and see several steps into the future. Impatience, on the other hand, only makes you look weak. It is a principal impediment to power.

Half of your mastery of power comes from what you do *not* do, what you do *not* allow yourself to get dragged into. For this skill you must learn to judge all things by what they cost you.

To be a master player you must be a master psychologist. ... An understanding of people's hidden motives is the single greatest piece of knowledge you can have in acquiring power.

Never trust anyone completely and study everyone, including friends and loved ones.

We now turn to some of the 48 laws of power:

NEVER OUTSHINE THE MASTER

Always make those above you feel comfortably superior. In your desire to please or impress them, do not go too far in displaying your talents or you might accomplish the opposite—inspire fear and insecurity.

Make your masters appear more brilliant than they are and you will attain the heights of power.

NEVER PUT TOO MUCH TRUST IN FRIENDS, LEARN HOW TO USE ENEMIES

Be wary of friends—they will betray you more quickly, for they are easily aroused to envy. They also become spoiled and tyrannical. But hire a former enemy and he will be more loyal than a friend, because he has more to prove. In fact, you have more to fear from friends than from enemies. If you have no enemies, find a way to make them.

WIN THROUGH YOUR ACTIONS, NEVER THROUGH ARGUMENT

Any momentary triumph you think you have gained through argument is really a Pyrrhic victory: The resentment and ill will you stir up is stronger and lasts longer than any momentary change of opinion. It is much more powerful to get others to agree with you through your actions, without saying a word. Demonstrate, do not explicate.

USE SELECTIVE HONESTY AND GENEROSITY TO DISARM YOUR VICTIM

One sincere and honest move will cover over dozens of dishonest ones. Open-hearted gestures of honesty and generosity bring down the guard of even the most suspicious people. Once your selective honesty opens a hole in their armor, you can deceive and manipulate them at will. A timely gift—a Trojan horse—will serve the same purpose.

POSE AS A FRIEND, WORK AS A SPY

Knowing about your rival is critical. Use spies to gather valuable information that will keep you a step ahead. Better still: Play the spy yourself. In polite social encounters, learn to probe. Ask indirect questions to get people to reveal their weaknesses and intentions. There is no occasion that is not an opportunity for artful spying.

DO NOT BUILD FORTRESSES TO PROTECT YOURSELF—ISOLATION IS DANGEROUS

The world is dangerous and enemies are everywhere—everyone has to protect themselves. A fortress seems the safest. But isolation exposes you to more dangers than it protects you from—it cuts you off from valuable information, it makes you conspicuous and an easy target. Better to circulate among people, find allies, and mingle. You are shielded from your enemies by the crowd.

PLAY THE PERFECT COURTIER

The perfect courtier thrives in a world where everything revolves around power and political dexterity. He has mastered the art of indirection; he flatters, yields to superiors, and asserts power over others in the most oblique and graceful manner. Learn and apply the laws of courtiership and there will be no limit to how far you can rise in the court.

PLAY ON PEOPLE'S NEED TO BELIEVE TO CREATE A CULTLIKE FOLLOWING

People have an overwhelming desire to believe in something. Become the focal point of such desire by offering them a cause, a new faith to follow. Keep your words vague but full of promise; emphasize enthusiasm over rationality and clear thinking. Give your new disciples rituals to perform, ask them to make sacrifices on your behalf. In the absence of organized religion and grand causes, your new belief system will bring you untold power.

CONTROL THE OPTIONS: GET OTHERS TO PLAY WITH THE CARDS YOU DEAL

The best deceptions are the ones that seem to give the other person a choice: Your victims feel they are in control, but are actually your puppets. Give people options that come out in your favor whichever one they choose. Force them to make choices between the lesser of two evils, both of which serve your purpose. Put them on the horns of a dilemma: They are gored wherever they turn.

PLAY TO PEOPLE'S FANTASIES

The truth is often avoided because it is ugly and unpleasant. Never appeal to truth and reality unless you are prepared for the anger that comes from disenchantment. Life is so harsh and distressing that people who can manufacture romance or conjure up fantasy are like oases in the desert: Everyone flocks to them. There is great power in tapping into the fantasies of the masses.

THINK AS YOU LIKE BUT BEHAVE LIKE OTHERS

If you make a show of going against the times, flaunting your unconventional ideas and unorthodox ways, people will think that you only want attention and that you look down upon them. They will find a way to punish you for making them feel inferior. It is far safer to blend in and nurture the common touch. Share your originality only with tolerant friends and those who are sure to appreciate your uniqueness.

WORK ON THE HEARTS AND MINDS OF OTHERS

Coercion creates a reaction that will eventually work against you. You must seduce others into wanting to move in your direction. A person you have seduced becomes your loyal pawn. And the way to seduce others is to operate on their individual psychologies and weaknesses. Soften up the resistant by working on their emotions, playing on what they hold dear and what they fear. Ignore the hearts and minds of others and they will grow to hate you.

DISARM AND INFURIATE WITH THE MIRROR EFFECT

The Mirror reflects reality, but it is also the perfect tool for deception: When you mirror your enemies, doing exactly as they do, they cannot figure out your strategy. The Mirror Effect mocks and humiliates them, making them overreact. By holding up a mirror to their psyches, you seduce them with the illusion that you share their values; by holding up a mirror to their actions, you teach them a lesson. Few can resist the power of the Mirror Effect.

NEVER APPEAR TOO PERFECT

Appearing better than others is always dangerous, but most dangerous of all is to appear to have no faults or weaknesses. Envy creates silent enemies. It is smart to occasionally display defects, and admit to harmless

*vices, in order to deflect envy and appear more
human and approachable. Only gods and the dead
can seem perfect with impunity.*

ASSUME FORMLESSNESS

*By taking a shape, by having a visible plan, you open
yourself to attack. Instead of taking a form for your
enemy to grasp, keep yourself adaptable and on the
move. Accept the fact that nothing is certain and no
law is fixed. The best way to protect yourself is to be as
fluid and formless as water; never bet on stability or
lasting order. Everything changes.*

Appendix D

Research for Virtual Worlds' Promotion of Oxytocin

Here we share a series of articles and websites and their links related to brain chemistry, sense of self, and virtual worlds, particularly with regard to oxytocin. The articles provide interesting suggestions about the physical impact of immersing in virtual worlds and how the mind melds into virtual worlds.

"Social Networking Affects Brains Like Falling in Love." http://www.fastcompany.com/magazine/147/doctor-love. html

"Brain Chemistry: The encoding of memory is enabled by dopamine production in the brain; the work of M. Koepp, et al. (1998) showed video games generate almost double the levels of dopamine experienced by humans at rest. Performance doubled as well." http://www.kauffman.org/education/try-gaming.aspx

"Applied Nothingness: Nothingness and Psycho-Social Systems." http://www.nothingnesstheory.com/Applied%20 Nothingness%20Theory2.htm

"The Health Benefits of Social Media." http://compukol.com/blogs/compukol/the-health-benefits-of-social-media/

"Brain Basis of Human Social Interaction: From Concepts to Brain Imaging." http://physrev.physiology.org/content/89/2/453.full

"Oxytocin—The Elixir of Trust." http://www.brainsexmatters.com/news.php

"What Will Uploading Ourselves yield, virtual empathy or narcissism? 15 January 2011. Drew 3000 blog at cached site http://webcache.com/search?q=cache:R1fqV2-na_8J:drew3000.net/2011/01/15/virtual-empathy/t&cd=2&hl=en&CT=clnk&gl=us

"Virtual-Reality Video Game to Help Burn Patients Play Their Way To Pain Relief." http://www.sciencedaily.com/releases/2008/03/080319152744.htm

"Brain Pain." http://www.sciencentral.com/articles/view.php3?type=article&article_id=218392308

"Easing Pain for Burn Victims using Virtual Reality." http://www.bbc.co.uk/news/health-12297569

"Does the Internet Make You Happy?" http://athinklab.com/2011/03/18/sxsw-panel-does-the-internet-make-you-happy/

Appendix E

On the Information Explosion

The historic industrial revolution resulted from advances in manufacturing technology and innovative ideas concerning organizing specialization in sets of distinctly separate but interrelated tasks. The tremendous advances in computer and communications technologies in the last decades of the 20th century enabled a mind-boggling information explosion that overwhelmed our ability to keep up. Fortunately, as the principal global means for virtual interactions, i.e., the Internet, has evolved along with brilliant search engine inventions, so has our ability to find the information we really need. This has also spawned the multifaceted phenomena of social electronic networks that are changing the mindsets of future generations. Who knows where this will all lead?! Some further thoughts on all this follow.

> See ReadWriteWeb, "How Humanity Created So Much Data and Computable Knowledge (Infographic) (http://www.readwriteweb.com/archives/ how_humanity_created_so_much_data_comput- able_knowl.php?utm_source=pulsenews&utm_ medium=referral&utm_campaign=Feed%3A+readwrit eweb+%28ReadWriteWeb%29):
>
> "[When] I first looked at the completed timeline," Wolfram writes, "the first thing that struck me was

how much two entities stood out in their contributions: ancient Babylon, and the United States government ... [It] is sobering to see how long the road
to where we are today has been. But it is exciting
to see how much further modern technology has
already made it possible for us to go."

We've written here for several years about the
explosion of data production that's beginning and
will be a major factor in determining the nature of
human civilization in the near-term. In terms of
sheer quantity, far more will be made measurable
in the next few years than has been instrumented
by any of the other developments on Wolfram's
timeline. Google's Marissa Mayer calls the coming
Internet of Things "bigger than Moore's law." Former
HP CEO Mark Hurd said in 2009: "more data will be
created in the next four years than in the history of
the planet." What will we do with all that data? That's
up to us as a society, but it's a good idea to see it
coming and look at it within a historical context
(Kirkpatrick, 2011).

For a fascinating view of historical milestones, embodied
in a "Timeline of Systematic Data and the Development of
Computable Knowledge," from 20,000 B.C. through the year
2010, refer to the Timeline infographic at http://www.wolfra-
malpha.com/docs/timeline/.

Also see "What is informatics," https://www.informatics.
illinois.edu/display/extra/What+is+informatics:

The ability to handle vast amounts of information
cheaply has changed the way we live. Advances
in computing power, the World Wide Web, search
engines, and large-scale collaborative initiatives
like Wikipedia have revolutionized the way knowledge is created and shared. We have new forms

of social interaction—from email, IM, and blogs to eBay, Facebook, and YouTube—and collaborative art and entertainment—from Limewire and podcasts to Guitar Hero and Second Life (refer to Appendix G for more on Second Life and other collaboration platforms). Information technology (IT) has become a ubiquitous, indispensable component of our everyday lives, helping—or hindering—us as we manage information, create knowledge, and make decisions.

Within the humanities, digital content is changing the way we visualize, present, understand, and experience history and literature. Within the fine arts, artists are using high-tech tools to construct virtual worlds, produce animations, and make music. Within the social, biological, and physical sciences, pattern analysis, data mining, visualization of massive data sets, and large-scale simulation of biological and physical processes, are enabling new discoveries and insights.

Also, the technology we are creating to handle the greater complexity and information explosion in our world is, in turn, changing our ever-malleable brain, and thus changing the way we interact and learn. See "Modern Technology Is Changing the Way Our Brains Work, Says Neuroscientist" at http://www.dailymail.co.uk/sciencetech/article-565207/Modern-technology-changing-way-brains-work-says-neuroscientist. html#ixzz1VceMMNSb

Appendix F

On the Deeper Impact of Virtual Worlds

Understanding the world from another's perspective of how the world works is influenced by many dynamic variables. Being able to immerse oneself into that perspective, experiencing firsthand the contextual experiences and emotional reactions that help to develop those perspectives, as well as to engage in an environment that is not predictable but is ever changing (enabling emergence similar to the physical world) is a significant method for understanding another's view of the world. 3D Immersive Virtual Learning Environments are powerful tools to enable this ability for understanding another's unique perspective. Through enabling presence and transference, in addition to emergence from complex adaptive systems, 3D Immersive Learning Environments (3DILEs) facilitate a merger of our physical world with the virtual, providing a physically safe environment that is able to impact us behaviorally, emotionally and psychologically in order to understand another's view of reality, as well as to learn and practice communication and collaboration skills in complex group dynamics.

Following are resources that expound upon the impact of this significant environment and its uses.

"Online Therapy Institute's Trainings Receive BACP Endorsement!" Online Therapy Institute, November 11, 2010, http://shar.es/0PMV8

"New Research on Use of Online Resources by Male Adult Survivors of Abuse," Online Therapy Institute, November 10, 2010, http://shar.es/0PMTq

"How We Work in Second Life," Online Therapy Institute, http://www.onlinetherapyinstitute.com/second-life/ (information about impact of avatar on individual)

"Coming Home: Transitional Online Post-deployment Soldier Support in Virtual Worlds," University of Southern California, Institute for Creative Technologies, 2009, http://www.cominghomecenter.org

Dr. Kevin M. Holloway, Project Manager, National Center for Telehealth and Technology, "Telehealth and Technology: Psychological Applications," YouTube, 2010 presentation, http://www.youtube.com/watch?v=Sqr_BygETSs

Not a Game: Inside Virtual Iraq: Read the story by Sue Halpern in *The New Yorker* (May 19, 2008): http://www.newyorker.com/reporting/2008/05/19/080519fa_fact_halpern. Scenarios from Virtual Iraq, a virtual-reality simulation used to treat veterans suffering from posttraumatic stress disorder. Virtual Iraq is adapted from the video game Full Spectrum Warrior (http://www.youtube.com/watch?v=R6kl2BuhKmM)

Snow Worlds involves pain management through immersion in virtual worlds. Read the article and see the video: http://www.sciencentral.com/video/2008/11/11/virtual-reality-helps-war-heroes-recover-from

Club One Island Weight Loss (http://www.youtube.com/watch?v=SswlXujVUxk). Club One Island is a virtual health world focused on changing behavior to improve lives. Club One Island combines cognitive behavioral, gaming, and social-networking elements with a rich sensory experience for the greatest impact on personal habits. Club One Island is the first of many immersive digital environments from Club One, one of the premier fitness club networks in California. Learn more about Club One Island at www.cluboneisland.com or http://www.Facebook.com/cluboneisland

Education in Virtual Worlds

1. http://atlanticuniv.academia.edu/NancyZingrone/ Blog/60860/Education-in-Virtual-Worlds
2. Georgia Public School Systems: NOBLE Virtual World for Students on OpenSim Platform
   a. http://www.forsyth.k12.ga.us/site/default.aspx?PageTy pe=3&ModuleInstanceID=8626&ViewID=047E6BE3-6D87-4130-8424-D8E4E9ED6C2A&RenderLoc=0&FlexD ataID=47550&PageID=1
   b. How a Georgia district built its grid, http:// www.hypergridbusiness.com/2012/03/ how-a-georgia-district-built-its-grid/

The Virtual Framework, a joint venture of the Department of Defense and Private Enterprise:

1. http://virtualworldframework.com/web/about.html; http:// v2.modsim.org/news/dod-virtual-worlds-framework-announced /
2. http://www.defensenews.com/arti-cle/20120127/TSJ01/301270004/ New-Tool-Aims-Slash-Costs-Creating-Virtual-Worlds
3. http://www.openaffairs.tv/2012/03/ himss12-richard-boyd-lockheed-martin/

Additional Publications

Journal of Virtual Worlds Research http://jvwresearch.org/
Best Practices in VWs http://previewpsych.org/BPD2.0.pdf
Virtual Human Interaction Lab publications http://vhil.stan-ford.edu/pubs/
Institute for Creative Technologies http://ict.usc.edu/

Books

Blascovich, Jim & Bailenson, Jeremy (2011) Infinite Reality: Avatars, Eternal Life, New Worlds, and the Dawn of the Virtual Revolution. New York, NY: William Morrow Publishers.

Kapp, Karl M. & O'Driscoll, Tony (2010) Learning in 3D: Adding a New Dimension to Enterprise Learning and Collaboration. Pfeiffer Publishers.

Aldrich, Clark (2009) Learning Online with Games, Simulations, and Virtual Worlds: Strategies for Online Instruction. Jossey-Bass Guides to Online Teaching and Learning, Clark Aldrich (Author). Visit Amazon's Clark Aldrich Page. Find all the books, read about the author, and more. See search results for this author. Are you an author? Learn about Author Central.

Reeves, Byron & Read, J. Leighton (2009) Total Engagement: Using Games and Virtual Worlds to Change the Way People Work and Businesses Compete. Harvard Business School Press.

McGonigal, Jane (2011) Reality Is Broken: Why Games Make Us Better and How They Can Change the World. Penguin Press HC, Jane McGonigal (Author). Visit Amazon's Jane McGonigal Page. Find all the books, read about the author, and more. See search results for this author. Are you an author? Learn about Author Central.

Ahn, S. J. (2011). Embodied experiences in immersive virtual environments: effects on pro-environmental attitude and behavior. Stanford University, Dissertation, (May), http://vhil.stanford.edu/pubs/2011/ahn-embodied-experiences.pdf

Bailenson, J.N., Yee, N., Blascovich, J., & Guadagno, R.E. (2008). Transformed social interaction in mediated interpersonal communication. In Konijn, E., Tanis, M., Utz, S., & Linden, A. (Eds.), Mediated Interpersonal Communication (pp. 77–99). Lawrence Erlbaum Associates.

Bartle, R. (2004). Pitfalls of virtual property. The Themis Group, http://www.themis-group.com

Bente, G., Rüggenberg, S. & Krämer, N.C. (2004). Social Presence and interpersonal trust in avatar-based, collaborative net-communications. In: Proceedings of the Seventh Annual International Workshop Presence 2004. UVP, Valencia (S.54-61), http://www.temple.edu/ispr/prev_conferences/proceedings/2004/Bente,%20Ruggenberg,%20Kramer.pdf

Blascovich, J., & Bailenson, J. (2011) Excerpt: the introduction. Infinite Reality: Avatars, Eternal Life, New Worlds, and the Dawn of the Virtual Revolution. New York, NY: William Morrow Publishers, http://www.infinitereality.org/book/introduction_excerpt.pdf

Blascovich, J., & McCall, C. (2011) Attitudes in virtual reality. Crano, W. D., Cooper, J.,& Forgas, J. P. (Eds.), The Psychology of Attitudes and Attitude Change. New York, NY: Psychology Press.

DeAngelis, T. (2009) Can Second Life therapy helps with autism? American Psychological Association, September, 40(8), p. 40, http://www.apa.org/monitor/2009/09/second-life.aspx

DeAngelis, T. (2009) Virtual healing. American Psychological Association, September, 40(8), p. 36, http://www.apa.org/monitor/2009/09/virtual-healing.aspx

Doyle, D. (2009) Embodied presence: the imaginary in virtual worlds. In: Embodiment and Performativity, Digital Arts and Culture 2009, Arts Computation Engineering, UC Irvine.

Gilbert, R. L. (2011) The P.R.O.S.E. (Psychological Research on Synthetic Environments) Project: Conducting In-World Psychological Research on 3D Virtual Worlds. Journal of Virtual Worlds Research, July, 4(1).

Gorini, A., Capideville, C. S., De Leo, G., Mantovani, F., & Riva, G. (2011) The Role of Immersion and Narrative in Mediated Presence: The Virtual Hospital Experience. Cyberpsychology, Behavior, and Social Networking, 14(3), 99–105.

Gorini, A., Gaggioli, A., Vigna, C., & Riva, G. (2008) A Second Life for ehealth: prospects for the use of 3-d virtual worlds in clinical psychology. Journal of Medical Internet Research, 10(3):e21, http://www.jmir.org/2008/3/e21/

Hoch, D.B., Watson, A.J., Linton, D.A., Bello, H. E., & Senelly, M., et al. (2012) The Feasibility and Impact of Delivering a Mind-Body Intervention in a Virtual World. PLoS ONE 7(3): e33843. doi:10.1371/journal.pone.0033843.

Hoffman, H. G. (2004) Virtual-Reality Therapy. Scientific American, August, 58–65.

Kennedy, R. S. (2012) The Virtual Meets the Real, Part 1: Virtual environments and games as therapeutic tools. NeuroscienceCME, January, 18(7), 1, http://www.neurosciencecme.com/email/2012/011712.htm

Lesgold, A. M. (2001) The Nature And Methods Of Learning By Doing. American Psychologist, 56(11), 964–973.

McKerlich, R., Anderson, T., Riis, M., & Eastman, B. (2011) Student perceptions of teaching presence, social presence and cognitive presence in a virtual world. Journal of Online Learning and Teaching, September, 7(3).

Mehdi, E., Nico, P., Julie, D., & Bernard, P. (2004) Modelling character emotion in an interactive virtual environment. In: proceedings of AISB 2004 .Symposium: Motion, Emotion and Cognition. Leeds, UK (29 March – 1 April).

Morie, J. F. (2009) Re-entry: online virtual worlds as a healing space for veterans. McDowall, I. E., & Dolinsky, M. (Eds.) The Engineering Reality of Virtual Reality 2009. In: proceedings of the SPIE, Volume 7238 (2009), pp. 72380C-72380C-9.

Morrison, R. (2009) Empathy from avatars: propositions for improving trust development in pseudo-social relationships with avatars. European Journal of Social Sciences, 12(2).

Rothbaum, B. O. (2006) Virtual reality in the treatment of psychiatric disorders. CNS Spectrums, January, 11:1, http://mbldown-loads.com/0106CNS_GuestEd.pdf

Salazar, J. (2005) Complex Systems Theory, Virtual Worlds & MMORPG's: Complexities Embodied. Changing Views: Worlds in Play, DiGRA Digital Library. June. Vancouver: University of Vancouver, http://www.digra.org/dl/display_html?chid=http://www.digra.org/dl/db/06276.36443.pdf

Salazar, J., Gakuin, T., & Kenkyuka, D. S. (2005) On the ontology of mmorpg beings: a theoretical model for research. In: proceedings of DiGRA 2005 Conference.

Seegert, A. (2009) 'Doing there' vs. 'being there': performing presence in interactive fiction. Journal of Gaming and Virtual Worlds, 1(1).

Vasalou, A., Joinson, A. N., Keynes, M., & Pitt, J. (2007) Constructing my online self: avatars that increase self-focused attention. Computers in Human Behavior, 7(2), 445-448.

Wiecha, J. et al. (2010) "Learning in a Virtual World: Experience With Using Second Life for Medical Education." Journal of Medical Internet Research, 12 (1). http://www.jmir.org/2010/1/e1/

Yee, N., Bailenson, J. N., & Ducheneaut, N. (2009) The Proteus Effect: implications of transformed digital self-representation on online and offline behavior. Communication Research, 36(2), April, 285–312.

Zheng, D., & Newgarden, K. (2012) Rethinking language learning: virtual worlds as a catalyst for change. international Journal of Learning and Media, 3(2), 13–36.

Appendix G

Web Collaboration, Workspace, Blog Platforms

Books

Richard Mansfield. *How to Do Everything with Second Life®*. New York: McGraw-Hill, 2008.

Sarah Robbins and Mark Bell. *Second Life for Dummies*. Hoboken, NJ: Wiley, 2008.

Michael Rymaszewski, Wagner James Au, Cory Ondrejka, and Richard Platel. *Second Life: The Official Guide*. Indianapolis, IN: Wiley, 2008.

Aimee Weber, Kimberly Rufer-Bach, and Richard Platel. *Creating Your World: The Official Guide to Advanced Content Creation for Second Life*. Indianapolis, IN: Wiley, 2008.

Second Life Uses

Second Life is only one of a host of platforms that are being used in similar ways for Education, Business, Collaboration, Art, and Learning Simulations to name a few. It is one of the oldest and, thus, has more research and resources that we will note here. Some other platforms often used by educators include Open Sim, Unity 3D, MOSES (the US Army's experimental platform for non-military researchers: http://sitearm. wordpress.com/2011/05/16/u-s-army-offers-moses-3d-web-system-for-non-army-researchers/), and World of Warcraft

(the educators' guild, Cognitive Dissonance is found here with over 300 members: http://cognitivedissonance.guildportal.com/ Guild.aspx?GuildID=228854&TabID=1927706. You can also find more information about practical application of WoW in school systems at their WoWinSchool wiki: http://wowin-school.pbworks.com/w/page/5268731/FrontPage.)

This website is an avenue for getting started in Second Life: http://secondlifegrid.net/

Second Life Tutorials

- **Tutorials:** http://secondlife.com/showcase/tutorials/
- **Mental Mentors:** http://wiki.secondlife.com/wiki/ Mental_Mentors_Page_3

Mixed Reality Intercultural Online Games in Second Life

There is information on this site in Second Life http://search. secondlife.com/web/search/?q=Mixed+Reality+Intercultural+O nline+Games&s=secondlife_com&m=N&lang=en-US). It seems to be along the lines of Alfred Hubler's changing views of reality through mixed reality states and virtual games. Hubler is affiliated with the University of Illinois Understanding Complexity (UIUC) effort and their Singapore Cultural Conflict Center.

SIETAR

Explore different cultures through nineteen game-based exercises in the SIETAR Intercultural Learning Space.

- **Second Life website:** http://slurl.com/secondlife/ Bluepill/226/202/67/
- **Sietar website:** http://www.sietar.org/

This may be the type of thing to enable a paradigm shift on a cultural level. Lots of universities are there with learning centers, research simulations, and online classrooms, e.g., http://secondlife.com/showcase/education/

Drawing Objects for Second Life

■ How does one create an avatar? http://www.archipelis. com/?gclid=CMKP7Ln-_5YCFQMCagodNgvaXQ

Social Scientist Collaboration in Second Life

■ The Social Simulation Research Lab is a library with 150+ resources (papers, websites, homepages, and references) of interest to social scientists in the virtual world. Dr. Aleks Krotoski (Second Life Avatar name: Mynci Gorky) has archived the research at http://mypages.surrey.ac.uk/ psp1ak/links.htm . Her current website : http://alekskroto-ski.com/ has even more related to her current research.

How Are Organizations Using the Second Life Platform?

Education and Nonprofits

The Second Life platform enhances distance learning, collaboration, demonstrations, and simulations. Early on, universities and libraries staked their claim in the virtual world and have since been pioneering uses of Second Life's unique tools to reach—and remain relevant to—a new generation. Nonprofits are also making their mark in-world and out. Both national-level and grassroots organizations conduct successful fund-raising events and information campaigns in Second Life that impact the lives of people everywhere.

■ **Second Life Wikia:** http://secondlife.wikia.com/index. php/Main_Page

Online Video Resources

■ Designing and Facilitating for Living Human Systems in Virtual Worlds http://youtu.be/pt11PHYxXSI
■ Infinite Reality: Avatars, Eternal Life and New Worlds (April 29, 2011) Jeremy Bailenson http://youtu.be/ lvGyK6vKIPE
  – Virtual Reality changes Real-Life Behavior http://youtu. be/3T9kfbcRrQ8
■ Jane McGonigal: Gaming can make a better world http:// youtu.be/dE1DuBesGYM
■ Eric Hackathorn (NOAA) Virtual Worlds in the Future http://youtu.be/YPadDVfSgDY
■ Eric Hackathorn YouTube Channel http://www.youtube. com/user/hackshaven
■ Behavior change, Club One Island in SL http://youtu.be/ SswlXujVUxk
■ Ann Cudworth_Data Analysis and Virtual Worlds http:// youtu.be/kiE1QqFjJv8
■ Monolith - Molecular Visualization for Second Life http:// youtu.be/3VXYz9W7y-M
■ SECOND LIFE - WUA VIRTUAL CAMPUS (Univ of Western Australia) http://youtu.be/0MyZ1WUY02s
■ Apollo 11 Tranquility Base Simulation in Second Life http://youtu.be/6qVvWOxzMDU
■ Airplane Tracking in Second Life http://youtu.be/ BMH2-rCkz3Y
■ MacArthur Foundation_shifts in way we learn http:// youtu.be/D6_U6jOKsG4
■ General brief overview for Govt http://youtu.be/ wKh8-QyL1Bs
  – **Military and Virtual Worlds:**

- Naval Undersea Warfare Center Division Newport http://youtu.be/iea8h_qcPOg; http://www.youtube. com/user/MaccusMccullough
- Related articles:
  - http://www.defense.gov/news/newsarticle. aspx?id=63874; http://www.informationweek. com/news/government/security/229500160
- HuffmanPrairieOmega_USAF http://youtu.be/ dQ8D4C1Q_ec
- Federal Virtual Worlds Challenge (sponsored by the Army):
  - Website with Videos of Winners: http://fvwc. army.mil/
  - Article: http://washingtontechnology.com/arti-cles/2011/06/02/army-names-winners-of-federal-virtual-worlds-contest.aspx
- Virtual World Advantages for Government Collaboration http://youtu.be/LX3xO1ojHmo
- **Federal Consortium for Virtual Worlds:**
  - 2012 http://www.ndu.edu/icollege/fcvw/2012post. html
  - 2011 http://vimeo.com/hackshaven/fcvw-overview ;
  - 2010 http://www.youtube.com/ watch?v=6qJxcCIUHL4
  - 2011 conference presentations http://www.ndu.edu/ icollege/fcvw/agendaD1.html
- **Federal Consortium for Virtual Worlds:**
  - Innovations in Virtual Worlds_FCVW 2010 Conference_Psychology, Health, CyberSecurity Education: http://youtu.be/fNEQz8kcCSA
  - Telehealth and Technology (T2) _Psychological applications : http://youtu.be/Sqr_BygETSs
  - Virtual Reality Therapy: Inside Virtual Iraq: http:// youtu.be/R6kl2BuhKmM
  - Chicoma3_Coming Home.org.wmv http://youtu.be/ Jd7otXMlWXM

- • Virtual Ability Island http://youtu.be/_pgWhsm56aA
- – Virtual Worlds as Green Spaces http://youtu.be/4woBlomg1x8
- ■ Transference and avatars_using virtual worlds to aid in physical therapy for burn victims http://youtu.be/jNIqyyypojg
- ■ 2008 Second Life Loyalist College Canadian Border Simulation http://youtu.be/PCUWcpVPtMM
- ■ 2011 Respiratory Ward HUD
  - – Loyola Marymount University in Second Life http://www.youtube.com/watch?v=RD5vkIFjRYk
- ■ 2011 Respiratory Ward HUD in Second Life - Imperial College London
  - – http://online.wsj.com/video/medical-training-in-second-life/9F96D4FB-AFF3-4D08-8F3F-E437AF63B974.html?mod=googlewsj ;
  - – http://www.youtube.com/watch?v=s1hOsiiHlyo ; http://www.youtube.com/watch?v=VhQ8MjdRq_4&feature=related
- ■ Science and Second Life: http://secondlife.com/destinations/science
- ■ Power Plant Orientation/Training using virtual world space http://www.youtube.com/watch?v=tSxonSQGhGg
- ■ The use of Second Life in teaching at Memorial University.
  - – http://www.delts.mun.ca/portal/index.php?SAID=132&Cat=%22Teaching_and_Technology%22
  - – Articles:
    - • http://www.universityaffairs.ca/memorials-second-life-shipyard.aspx ;
    - • http://community.secondlife.com/t5/Learning-Inworld-General/Memorial-University-is-Nationally-Recognized-in-Canada-for/ba-p/643492 ;

Explanation of importance of 3Di

- **The singularity:** http://www.youtube.com/watch?v=KdVcNy4cbiY
- **Webvolution:** http://www.youtube.com/watch?v=-cZTdFTZV5Q

YouTube Explanations of Second Life

- **Excellent overview:** Touches on personal tailoring of your interactions but emphasizes business applications to create collaborators and sell your company's products: http://www.youtube.com/watch?v=8NOHRJB9uyI&feature=related
- **Good general information:** Virtual Social Worlds and the Future of Learning: http://www.youtube.com/watch?v=O2jY4UkPbAc
- **Funny introductions**
  - **IBM:** http://www.youtube.com/watch?v=5ly4LIxzGZM
  - **For science and more:** http://www.youtube.com/watch?v=MRmtd4wm1RI
- **Tour of Second Life:** http://www.youtube.com/watch?v=VeOGOhvu4gM&feature=related
- **Introduction for Businesses:** http://www.youtube.com/watch?v=s0rqJtg7F5c
  - **Sunday *Times* example:** http://www.youtube.com/watch?v=nWE1FTgp1G0
  - **General introduction, including business:** http://www.youtube.com/watch?v=b72CvvMuD6Q
  - **Virtual workplaces** (great example of corporate collaboration): http://www.youtube.com/watch?v=Zy1aOGxe2Is
- **Real-life companies in Second Life**
  - **Globaling Business Island:** http://www.youtube.com/watch?v=8woeDWZggFI&feature=related
  - **Overview:** http://www.youtube.com/watch?v=fE_6Xv_rEt4&feature=related

- **Dell:** www.dell.com/secondlife
- **World Bank:** http://www.youtube.com/
  watch?v=YpbuW-s-oNI&feature=related
- **Organizations in Second Life**
  - **Institutional Use of Second Life:** http://www.
    youtube.com/watch?v=zkIoiCqoeEI
  - **Enterprise applications of Second Life:** http://
    www.youtube.com/watch?v=hmKaR6xUMAg
  - **NASA CoLabs's Second Life Mission: (http://
    www.youtube.com/watch?v=kr3vXuxEPB8)**
  - **NASA History through Second Life:** http://
    www.youtube.com/watch?v=fu2hvrMjgu0
- **Press Conferences in Second Life**
  - **Sun Microsystems:** http://www.youtube.com/
    watch?v=imauSAHvehU
  - **Mixed reality meeting:** http://www.youtube.com/
    watch?v=EIeUZ2njcDs
■ **Intro to Second Life for organizations**
  - **Introduction to Second Life (Historical View):**
    http://www.youtube.com/watch?v=b72CvvMuD6Q
  - **CISCO in Second Life:** http://www.youtube.com/
    watch?v=SLwtmOe4hzo&NR=1
  - **Virtual Greetings from John Chambers in Second
    Life:** http://www.youtube.com/watch?v=-FInj38liTw
■ **Education in Second Life**
  - **Gamer audience:** http://www.youtube.com/
    watch?v=qOFU9oUF2HA
  - **Virtual learning (lecture format):**
    http://www.youtube.com/watch?v=O2jY4UkPbAc
  - **Science learning opportunities in Second Life:**
    http://www.youtube.com/watch?v=EfsSGBraUhc
  - **Training simulations in Second Life:** http://www.
    youtube.com/watch?v=DJTzNSV8pb0
■ **Education, Research, Classes**

- **International Society for Technology in Education:** http://www.youtube.com/watch?v=aP137QgYKvQ
- **SIMs (Simulations)**
  - **Physics:** http://www.youtube.com/watch?v=5CzpB6Q2TNo
  - **Universities:** Many universities have simulation sites in second life.
    - **Princeton:** http://www.youtube.com/watch?v=g0_FEjjV-fA
    - **Ohio University SL campus:** http://www.youtube.com/watch?v=aFuNFRie8wA
    - **North Carolina State:** http://www.youtube.com/watch?v=CWfvqkkk0yM
- **Nonprofits**
  - **CRY in Second Life:** http://www.youtube.com/watch?v=P8qMXlFmnRY
  - **Dancing at Nonprofit Commons:** http://www.youtube.com/watch?v=ytT-toBfEyc
  - **Nonprofit Commons** (film made pre- corporate boom): http://www.youtube.com/watch?v=hqsmZRnvpnY

Articles about Second Life Uses

The following articles relate to what's been envisioned about Second Life. Will it be a major part of the change our global society is currently experiencing? Will the 3D Web be the next evolution?

- **What's next in Second Life?** "Second Life: It's Not a Game," CNNMoney, January 23, 2007: http://money.cnn.com/2007/01/22/magazines/fortune/whatsnext_secondlife.fortune/index.htm
- **Business in Second Life**

- "You in Back, Yes You, the Half-Lynx," Los Angeles Times, May 10, 2008: http://articles.latimes.com/2008/may/10/business/fi-secondlife10
- "BITS; Second Life for Corporations," New York Times, April 7, 2008: http://query.nytimes.com/gst/fullpage.html?res=9B06E3DE1F3FF934A35757C0A96E9C8B63

■ **Virtual world for testing out ideas:** "Second Life's Virtual World Attracts W Hotels, American Apparel and Other Corporations," August 7, 2006: http://www.businessweek.com/innovate/NussbaumOnDesign/archives/2006/08/second_lifes_vi.html

■ **Slides of Second Life and corporations, nonprofits, education, art, media, 3D Web**
- "An Introduction to Virtual Worlds," Slideshare: http://www.slideshare.net/ialja/virtual-worlds-introduction-second-life-and-beyond
- "Arvetica: Second Life for Businesses," Slideshare: http://www.slideshare.net/Alex.Osterwalder/arvetica-second-life-for-businesses-introduction
- "Second Life Study," Slideshare: http://www.slideshare.net/joestu/second-life-study-swot-analysis-corporations-vs-experts

■ **Virtual worlds and simulations for medical and psychiatry treatments online:** "The Online Doctor Will See You Now," New Scientist, November 8, 2008: http://www.newscientist.com/article/mg20026816.000-the-online-doctor-will-see-you-now.html?full=true&print=true

■ **Virtual companies legal in Vermont:** Second Life Pros: http://www.secondlifepros.com/virtual-corporations-in-vermont/

■ **Second Life general information:** "Second Life—What You Should Know before Your Corporation Goes There," Web Strategy, May 19, 2007: http://www.web-strategist.com/blog/2007/05/19/second-life-what-you-should-know-before-your-corporation-goes-there/

■ **Virtual worlds insight:** Blog on brand, education, and collaboration in virtual worlds: http://dusanwriter.com/?gcl id=CPfQ9uS1nJcCFQSPFQodMBKl-Q

Semantic Web

■ **Explanation of the semantic web:** Kate Greene, "The Semantic Web Goes Mainstream," *Technology Review*, October 29, 2007: http://www.technologyreview.com/computing/19627/?a=f

■ **Semantic metadata review:** "Semantic Meta Data for Enterprise Information Integration," Information Management, July 1, 2003: http://www.dmreview.com/issues/20030701/6962-1.html

■ **Data integration:** "Oracle and Silver Creek Systems," Oracle, http://www.silvercreeksystems.com/?gclid=CJzW4Z rsyJYCFQt4HgodVm2GzQ

■ **A list of video explanations of the Semantic Web:** http://video.google.com/video search?hl=en&q=semantic+software&um=1&ie= UTF-8&sa=X&oi=video_result_group&resnum=4&ct=title#

■ **List of links of limited comparisons:** WSindex, http://www.wsindex.org/Companies/Semantic_Web/index.html

■ **Semantic normalization:** "Making Sense out of Health Data," Intel, http://software.intel.com/en-us/blogs/2008/09/23/semantic-normalization-making-sense-out-of-health-data/

■ **Google search of semantic + decision-making software:** http://www.google.com/m?eosr=on&q=semantic%2 Bdecision+making+software&start=10&sa=N&mrestrict=xh tml

■ Semantic networks as a model for mapping the rich interconnectedness of all things:
  – **Semantic Research:** http://www.semanticresearch. com/

- **i2 TextChart:** http://www.i2inc.com/
  products/textchart/?_kk=semantic%20
  software&_kt=818f0d32-d69f-40d4-9e1a-10b8d89a1df4
- **Software Abstractions Blog:** http://blog.softwareab-
  stractions.com/the_software_abstractions/semantic_web/
- **Tech entrepreneur forum**, Northeastern University
  School of Technological Entrepreneurship: http://www.
  entretechforum.org/mm_Feb19_2008.htm
- **Lexical semantics:** "Lexical Semantics for Software
  Requirements Engineering: A Corpus-Based Approach,
  ingentaconnect: http://www.ingentaconnect.com/content/
  rodopi/lang/2007/00000062/00000001/art00020;jsessionid
  =3dw0ptt2hgwpf.alexandra
- **2010 Semantic Technology Conference:** http://www.
  semantic-conference.com/sponsors/
- **The semantic web mobile edition:** ZDNet: http://blogs.
  zdnet.com/semantic-web/wp-mobile.php
- **The Software Abstractions Blog:** semantic
  web articles: http://blog.softwareabstractions.com/
  the_software_abstractions/semantic_web/

Website Workplace Software

- **Web conferencing reviews and articles:** Think Of
  It: http://thinkofit.com/webconf/wcreview.htm
- **GoToMeeting:** https://www1.gotomeeting.com/
- **Think out of the box:** Collanos: http://www.collanos.com/
- **WebEx weboffice for nonprofits:** Nonprofit intranet,
  calendar software, online database, weboffice: http://
  www.weboffice.com/EN/Pricing/NonProfit/
- **WebEx weboffice:** http://www.weboffice.com/EN/Home/
  default.asp
- **Facilitate Pro Web Meeting Software:** http://www.
  facilitate.com/introduction.html?src=google&gclid=CPCxof
  n4yJYCFQxxHgodGGYmxw

- **Online project management, ProjectSpaces:** http://www.projectspaces.com/section/features/
- **Online collaboration web:** based on IBM/Lotus Quickplace Quickr software hosting: http://www.project-lounge.com/
- **Planzone project management online software:** http:planzone.com
- **Free wiki/ business, PBworks:** http://pbwiki.com/
- **MyOffice:** http://www.myoffice.net/
- **iCohere: Collaboration software and consulting for organizational learning, innovation, and community, iCohere:** http://www.icohere.com/
- **GroupSwim:** http://groupswim.com/
- **eUnify Networks collaboration tools:** http://www.eunify.net/
- **eTouch SamePage, the enterprise wiki:** http://www.etouch.net/home/
- **Online collaborative work environments:** Think of It: http://thinkofit.com/webconf/workspaces.htm
- **Central Desktop:** http://www.centraldesktop.com/l?sr=af_spjb4tpx3wwknvc9i98n

Nonprofit Fundraising Software

- **GiftWorks fundraising software:** http://blog.missionresearch.com/2006/06/volunteers_sema.html

References

Abrams, Lisa C. 2003. "Nurturing Interpersonal Trust in Knowledge-Sharing Networks." *Academy of Management Executive* 17(4):64–77.

Abuzaakouk, Asma, and Rob Creekmore. 2008. "Knowledge Harvesting for Organizational Learning." Presentation at Organizational Learning Technical Exchange Meeting I. The MITRE Corporation, January 14.

Ackoff, Russell L. 2004. "Transforming the Systems Movement." *The Systems Thinker* 15(8): Pegasus Communications, http://www.pegasuscom.com/tstpage.html

Ahuja, Anjana. 2010. "The Natural Selection of Leaders." *New Scientist*, September 4, 28–29.

Ancona, Deborah, and Henrik Bresman. 2007. *X-Teams: How to Build Teams That Lead, Innovate, and Succeed*. Boston, MA: Harvard Business School Press.

Ariely, Dan. 2009. *Predictably Irrational—The Hidden Forces That Shape Our Decisions*. Revised and expanded edition. New York: Harper Collins.

Argyris, Chris, and Donald Schön. 1974. *Theory in Practice: Increasing Professional Effectiveness*. San Francisco: Jossey-Bass.

Argyris, Chris, and Donald Schon. 1978. *Organizational Learning: A Theory Of Action Perspective*. Reading, MA: Addison-Wesley.

Argyris, Chris, Robert Putnam, and Diana McLain Smith. 1985. *Action Science: Concepts, Methods and Skills for Research and Intervention*. San Francisco: Jossey-Bass.

Argyris, Chris. 1991. "Teaching Smart People How to Learn." *Harvard Business Review* May–June:99–109.

Argyris, Chris. 1994. "Good Communication That Blocks Learning." Reprint 94401. *Harvard Business Review* July–August:43–53.

Ashby, W. Ross 1956. *An Introduction to Cybernetics.* New York: John Wiley.

Ashby, W. Ross 1958. "Requisite Variety and Implications for Control of Complex Systems." *Cybernetica* 1(2):83–99.

Axelrod, Robert, and Michael D. Cohen. 1991. *Harnessing Complexity—Organizational Implications of a Scientific Frontier.* New York: The Free Press.

Bacon, Terry R., and Karen I. Spear. 2003. *Adaptive Coaching: The Art and Practice of a Client-Centered Approach to Performance Improvement.* Palo Alto, CA: Consulting Psychologists Press.

Barch, John A., and Ezequiel Morsella. 2008. The Unconscious Mind. *Perspectives on Psychological Science* 3:73–79. doi:10.1111/j.1745-6916.2008.00064.x. http://pps.sagepub.com/content/3/1/73.full

Barfield, Becky. 2008. "The Impact of Emotion on Organizational Learning." Presentation at Organizational Learning Technical Exchange Meeting I. The MITRE Corporation, January 14.

Bar-Yam, Yaneer. 2002. "Complexity Rising: From Human Beings to Human Civilization, a Complexity Profile." New England Complex Systems Institute, Cambridge, MA. Encyclopedia of Life Support Systems (EOLSS, New York: Oxford University Press).

Bar-Yam, Yaneer. 2004. *Making Things Work: Solving Complex Problems in a Complex World.* Cambridge, MA: New England Complex Systems Institute.

Baskerville, Tracey A., and Alison J. Douglas. 2010. Dopamine and Oxytocin Interactions Underlying Behaviors: Potential Contributions to Behavioral Disorders. *CNS Neuroscience & Therapeutics 16(3):e92–e123.* Article first published online May 6, 2010. doi:10.1111/j.1755-5949.2010.00154.x. http://onlinelibrary.wiley.com/doi/10.1111/j.1755-5949.2010.00154.x/full

Beinhocker, Eric D. 2006. *The Origin of Wealth—The Radical Remaking of Economics and What It Means for Business and Society.* Boston: Harvard Business School Press.

Bennett, Drake. 2009. "Dark Green—A Scientist Argues that the Natural World Isn't Benevolent and Sustaining: It's Bent on Self-Destruction." *The Boston Globe,* January 11, C1–C2.

Bertalanffy, Karl Ludwig von. 1968. *General Systems Theory: Foundations, Development, Applications.* New York: George Braziller.

Boardman, John, and Brian Sauser. 2008. *Systems Thinking—Coping with 21st Century Problems*. Boca Raton, FL: CRC Press.

Boehm, Barry, Ricardo Valerdi, and Eric Honour. 2008. "The ROI of Systems Engineering: Some Quantitative Results for Software-Intensive Systems," *Systems Engineering* 11(3), 221–234.

Booker, Lashon B., and Gary W. Strong. 2008. "Using Topic Analysis to Compute Identity Group Attributes." pp. 249–258 in *Social Computing, Behavioral Modeling, and Prediction*, edited by Huan Liu, John J. Salerno, and Michael J. Young. New York: Springer.

Brafman, Ori, and Rod A. Beckstrom. 2006. *The Starfish and the Spider: The Unstoppable Power of Leaderless Organizations*. London: Penguin Books.

Brooks, David. 2011. *The Social Animal—The Hidden Sources of Love, Character, and Achievement*. New York: Random House.

Brown, Steven, and Lawrence M. Parsons. 2008. "So You Think You Can Dance? PET Scans Reveal Your Brain's Inner Choreography—Recent Brain-Imaging Studies Reveal Some of the Complex Neural Choreography Behind Our Ability to Dance." *Scientific American Magazine*, June 16.

Buchanan, Mark. 2008. "Why Complex Systems Do Better without Us." *New Scientist*, August 6, 28–31.

Buckley, Kerry. 2008. "Organizational Learning: The MITRE Experience." Presentation at Organizational Learning Technical Exchange Meeting I. The MITRE Corporation. January 14.

Buckley, Kerry et al. 2009. "Collaboration in the National Security Arena: Myths and Reality. What Science and Experience Can Contribute to Its Success." Collaboration White Paper. Topical Strategic Multi-Layer Assessment (SMA) Multi-Agency/Multi-Disciplinary White Papers in Support of Counter-Terrorism and Counter-WMD.

Burt, David N. 2001. "Institutional Trust." 86th Annual International Conference Proceedings. http://www.ism.ws/pubs/proceedings/confproceedingsdetail.cfm?ItemNumber=11800

Busch, Peter. 2008. *Tacit Knowledge in Organizational Learning*. Hershey, PA: IGI Global.

Calmes, Jackie. 2010. "Invoking the Oil Crisis, Obama Lauds Clean Energy." *New York Times*, May 26, http://green.blogs.nytimes.com/2010/05/26/invoking-the-oil-crisis-obama-lauds-clean-energy/

Carey, Benedict. 2008. "Tolerance Over Race Can Spread, Studies Find." *New York Times*, October 7.

Carroll, John S. 2008. "Organizational Learning Themes." Presentation at Organizational Learning Technical Exchange Meeting I. The MITRE Corporation. January 14.

Castka, P., C. J. Bamber, J. M. Sharp, and P. Belohoubeck. 2001. "Factors Affecting Successful Implementation of High-Performance Teams." *Team Performance Management* 7(7–8): 123–134.

Castronova, Edward. 2003. "Theory of the Avatar." CESifo Working Paper Series No. 863. Indiana University–Bloomington, Department of Telecommunications. CESifo (Center for Economic Studies and Ifo Institute for Economic Research), http://papers.ssrn.com/sol3/papers.cfm?abstract_id=385103

Cathcart, Rebecca. 2008. "Woman Pleads Not Guilty in Internet Suicide." Technology section. *The New York Times*, June 17.

Chesbrough, Henry W. 2003. "The Era of Open Innovation." *MIT Sloan Management Review* 44(3): 7.

Chown, Marcus. 2007. "Equation Can Spot a Failing Neighbourhood." *New Scientist* 2628: 8.

Cook, Timothy E., and Paul Gronke. 2001. "The Dimensions of Institutional Trust: How Distinct Is Public Confidence in the Media?" Annual Meeting of the Midwest Political Science Association, Chicago, April. http://www.reed.edu/~gronkep/papers.html

Covey, Stephan R. 2004. *The 8th Habit—From Effectiveness to Greatness*. New York: Free Press.

Creekmore, Ingram R. 2008. "Integrated Project Team (IPT) Start-up Guide." The MITRE Corporation, October.

Damasio, Antonio. 2005. *Decartes' Error: Emotion, Reasons, and the Human Brian*. New York: Penguin.

Damasio, Antonio R. 1994. *Descartes' Error: Emotion, Reason, and the Human Brain*. New York: G. P. Putnam's Sons.

DeCarlo, Douglas. 2004. *eXtreme Project Management: Using Leadership, Principles, and Tools to Deliver Value in the Face of Volatility*. San Francisco, CA: Jossey-Bass.

Demasio, Antonio R. 1994. *Descartes' Error: Emotion, Reason, and the Human Brain*. New York: G. P. Putnam's Sons.

Denning, Stephen. 2005. *The Leader's Guide to Storytelling: Mastering the Art and Discipline of Business Narrative*. New York: Jossey-Bass.

Detweiler, Karen. 2008. "Tacit Knowledge in the Workplace." Presentation at Organizational Learning Technical Exchange Meeting I. The MITRE Corporation, January 14.

DoD-AT&L (Department of Defense, Office of the Under Secretary of Defense for Acquisition, Technology, and Logistics). 2009. "Understanding Human Dynamics." Report of the Defense Science Board Task Force on Understanding Human Dynamics. Washington, DC: AT&L.

Douglas, Kate. 2007. "The Subconscious Mind: Your Unsung Hero." *New Scientist* 2632, December 1.

Dunn, Jennifer R., and Maurice E. Schweitzer. 2005. "Feeling and Believing: The Influence of Emotion on Trust." *Journal of Personality and Social Psychology* 88(5):736–748.

Economist. 2010a. B. G., "The 24-hour Athenian Democracy." Blog posting by B.G. *The Economist*, December 8, http://www.economist.com/blogs/babbage/2010/12/more_wikileaks

Economist. 2010b. "Unpluggable: How WikiLeaks Embarrassed and Enraged America Gripped the Public and Rewrote the Rules of Diplomacy," and "Even Those Who Back More Disclosure Should Hesitate before Condoning WikiLeaks' Torrent of E-mails." *The Economist*, December 2, http://www.economist.com/node/17633606 and http://www.economist.com/node/17629833/comments?page=3

Edmonson, Amy. 1999. "Psychological Safety and Learning Behavior in Work Teams." *Administrative Science Quarterly* 44(2): 350–383.

Edmondson, Amy C. 2008. "The Competitive Imperative of Learning." *Harvard Business Review,* July–August, 60–67.

Else, Liz. 2010. Wake Up and Smell the Apocalypse." *New Scientist*, August 28–September 3, 28–29.

Fearing, F. 1954. "An Examination of the Conceptions of Benjamin Whorf in the Light of Theories of Perception and Cognition." *American Anthropologist* 56.

Fine, Aubrey H. 2006. *Handbook on Animal-Assisted Therapy— Second Edition: Theoretical Foundations and Guidelines for Practice.* New York: Academic Press.

Fineman, S. 1993. *Emotion in Organizations.* London: Sage.

Fisher, Denise. 2005. "Falling in Love: The Chemistry of the First Breastfeed." *Health*, March 30. http://www.health-e-learning.com/resources/articles/34-falling-in-love

Fisher, Richard. 2006. "Why Altruism Paid Off for Our Ancestors." NewScientist.com news service. December 7, 2006 (vol 314, p. 1569).

Flaherty, James. 2010. *Coaching Evoking Excellence and Others*, 3rd ed. Burlington, MA: Butterworth-Heinemann.

Frith, Chris. 2008. "No One Really Uses Reason." *New Scientist* 2666, July 23.

Fukuyama, Francis. 1992. *The End of History and the Last Man*. New York: Avon. 1992 (p.19).

Garvin, David A., Amy C. Edmondson, and Francesca Gino. 2008. "Is Yours a Learning Organization?" *Harvard Business Review*, March, 109–116.

Gawande, Atul. 2007. *Better: A Surgeon's Notes on Performance*. New York: Picador.

Gazzaniga, Michael. 2007. University of California, Santa Barbara, Department of Psychology. "Brains, Minds, and Social Process." Carnegie Institution for Science Lecture, October 11, http://carnegiescience.edu/events/lectures/brains_minds_and_social_process

Gharajedaghi, Jamshid. 1999. *Systems Thinking: Managing Chaos and Complexity: A Platform for Designing Business Architecture*. Boston: Butterworth Heinemann.

Giles, Jim. 2008. "Our Psychology Helps Politicians Bend the Truth." *New Scientist,* October 10. http://www.newscientist.com/article/mg20026774.400

Gill, Kathy. 2011. "Deconstructing a Political Poll: How to Determine If a Political Poll Is Valid or Newsworthy." About.com, US Politics, http://uspolitics.about.com/od/campaignpolls/a/deconstructpoll.htm

Gladwell, Malcolm. 2002. *The Tipping Point: How Little Things Can Make a Big Difference*. Newport Beach, CA: Back Bay Books.

Goleman, Daniel. 1995. *Emotional Intelligence: 10th Anniversary Edition. Why It Can Matter More Than IQ*. New York: Bantam.

Goleman, Daniel, Richard E. Boyatzis, and Annie McKee. 2002. *Primal Leadership: Learning to Lead with Emotional Intelligence*. Boston, MA: Harvard Business School Publishing.

Goode, Harry, and Robert E. Machol. 1957. *Systems Engineering: An Introduction to the Design of Large-Scale Systems.* New York: McGraw-Hill.

Google. 2008. "Lively" 3-D virtual reality website (beta), http://www.lively.com/html/landing.html

Greene, Robert. 2000. *The 48 Laws of Power.* New York: Penguin Books.

Greenspan, Stanley I., and Stuart G. Shanker. 2004. *The First Idea— How Symbols, Language, and Intelligence Evolved from Our Primate Ancestors.* Boston: Da Capo Press.

Hammonds, Keith H. 2007. "How Google Grows ... and Grows ... and Grows." Fast Company, http://www.fastcompany.com/magazine/69/google.html

Handy, Charles 1995. "Trust and the Virtual Organization." *Harvard Business Review* 73(3): 9.

Health E-Learning. 2005. "Falling in Love: The Chemistry of the First Breastfeed." Health E-learning Online Education. March 30, http://www.health-e-learning.com/resources/articles/34-falling-in-love

Heath, Chip, and Dan Heath. 2005. *Made to Stick: Why Some Ideas Survive and Others Die.* New York: Random House.

Heath, Chip, and Dan Heath. 2007. *Made to Stick: Why Some Ideas Survive and Others Die.* New York: Random House.

Highfield, Roger. 2008. "Scientists Find 'Law of War' That Predicts Attacks." *London Daily Telegraph.* June 28.

Hock, Dee. 1998. "The Chaordic Organization: Out of Control and into Order," http://www.ki-net.co.uk/graphics/Dee%20Hock%20-%20The%20Chaordic%20Organization.pdf

Hock, Dee. 1999. *Birth of the Chaordic Age.* San Francisco: Berrett-Khoehler Publishers; summary Innervention website, 2010, http://www.innervention.nl/page21/page21.html

Hock, Dee. 2000. "The Art of Chaordic Leadership." *Leader to Leader* 15, http://www.leadertoleader.org/knowledgecenter/journal.aspx?ArticleID=62

Hubler, Alfred, and Vadas Gintautas. 2008. "Experimental Evidence for Mixed Reality States." 8th Understanding Complex Systems Symposium. University of Illinois, Champaign-Urbana, May 12.

Hughes, Larry. 2008. "On Learning: The Future of Air Force Education and Training." Presentation at Organizational Learning Technical Exchange Meeting II. The MITRE

Corporation. April 22. This information is in the public domain and is available at the following URLs: http://www.aetc.af.mil/search/generalsearch.asp?q=MyBase; http://www.aetc.af.mil/shared/media/document/AFD-081216-008.pdf (e.g., see Chart 6 for Figure 4.2).

Hybertson, Duane. 2009. *Model-Oriented Systems Engineering Science—A Unifying Framework for Traditional and Complex Systems*. Boca Raton, FL: Auerbach.

Iacono, C. Suzanne, and Suzanne Weisband. 1997. "Developing Trust in Virtual Teams." Proceedings of the 30th Annual Hawaii International Conference on System Sciences.

ICAO (International Civil Aviation Organization). 1996. "VHF Digital Link (VDL) TDMA Mode (Mode 3)—Standards and Recommended Practices—Draft." Appendix D to the Report on Agenda Item 4. AMCP/4-WP/70, April 4.

INSIGHT. 2008. "Special Feature—Systems Science: Deepening Our Understanding of the Theory and Practice of Systems Engineering." *INCOSE INSIGHT* 11 (January).

Jameson, Rob. 2008. "The Blunders that Led to the Banking Crisis." *New Scientist* 2675.

Johnson, Carolyn Y. 2010. "Group IQ: What Makes One Team of People Smarter Than Another? A New Field of Research Finds Surprising Answers." *Boston Globe*, December 19, K1–K2.

Kagan, Robert. 2008. *The Return of History and the End of Dreams*. New York: Alfred A. Knopf.

Kahan, Seth. 2006. "The Power of Storytelling to JumpStart Collaboration." *Journal for Quality & Participation* 29(1): 23–25.

Khamsi, Roxanne. 2007. "Impaired Emotional Processing Affects Moral Judgments." *NewScientist* 13, http://www.newscientist.com/article/dn11433-impaired-emotional-processing-affects-moral-judgements.html?full=true

Kirkpatrick, Marshall. 2011. "How Humanity Created So Much Data and Computable Knowledge (Infographic)". ReadWriteWeb, August 19, 2011, http://www.readwriteweb.com/archives/how_humanity_created_so_much_data_computable_knowl.php

Klein, Naomi. 2008. *The Shock Doctrine: The Rise of Disaster Capitalism*. New York: Picador.

Kleiner, Art. 2005. "Karen Stephenson's Quantum Theory of Trust." *strategy+business issue* 29, 11, October 2002, http:www.strategy-business.com. content/the creative mind. Reprint 02406.

1-14; also, Issue 8. Fieldnotes: A Newsletter of the Shambhala Institute. January, 2005, 1-9, http://www.shambhalainstitute. org/Fieldnotes/Issue8/I8_Kleiner.pdf

Klein, Harold, and William Newman. 1980. How to Use SPIRE: A Systematic Procedure for Identifying Relevant Environments for Strategic Planning. *Journal of Business Strategy.* 1-1: 32–45.

Knowles, Richard N. 2002. *The Leadership Dance—Pathways to Extraordinary Organizational Effectiveness* (3rd edition). Niagara Falls, NY: The Center for Self-Organizing Leadership.

Lancaster, Lynne C., and David Stillman. 2002. *When Generations Collide: Traditionalists, Baby Boomers, Generation Xers, Millennials. Who They Are. Why They Clash. How to Solve the Generational Puzzle at Work.* New York: Collins Business, HarperCollins Publishers.

Lawton, Graham. 2011. The Grand Delusion: Blind to Bias. *New Scientist*, 17 May 37–38.

Lencioni, Patrick. 2002. *The Five Dysfunctions of a Team: A Leadership Fable.* San Francisco: Jossey-Bass. 2002.

Leonard, Dorothy, and Walter Swap. 1999. *When Sparks Fly: Igniting Creativity in Groups.* Boston, MA: Harvard Business School Press.

Levin, Daniel Z., Rob Cross, Lisa C. Abrams, and Eric L. Lesser. 2002. "Trust and Knowledge Sharing: A Critical Combination." IBM Institute for Knowledge-Based Organizations, October 2002, http://www.935ibm.com/services/nZ/igs/pdf/g510-1693-OO-CPOV-trust-and-knowledge-sharing.pdf

MacKenzie, Debora. 2008a. "Will a Pandemic Bring Down Civilization?" *New Scientist* 2650:28–31.

MacKenzie, Debora. 2008b. "Why the Demise of Civilisation May Be Inevitable," *New Scientist* 2650:32–35.

Magnuson, Stew. 2008. "To Heal Psychological Trauma, Troops Relive War in Virtual Reality." *National Defense*, December. http://www.nationaldefensemagazine.org/archive/2008/December/Pages/ToHealPsychologicalTrauma,TroopsReliveWar inVirtualReality.aspx.

Maier, Mark W., and Eberhardt Rechtin. 2009. *The Art of System Architecting* (3rd edition, 395–408). Boca Raton, FL: CRC Press.

Malone, David. 2007. "Are We Still Addicted to Certainty?" *NewScientist* 2615, August 4.

Malone, Thomas W. 2010. "Group Collective Intelligence Predicts Group Performance in Many Situations," October, http://www.outlookseries.com/A0999/Science/3983_Thomas_W._Malone_MIT_Group_Collective_Intelligence_Predicts_Group_Performance_Thomas_W._Malone.htm

Malone, Thomas W. 2004. *The Future of Work*. Boston, MA: Harvard Business School Press.

Marshall, Michael. 2010. "Sparks Fly over Origin of Altruism." *New Scientist*, October 2, 8–9.

McCarter, Beverly Gay, and Brian E. White. 2007. "Collaboration/Cooperation in Sharing and Utilizing Net-Centric Information," Conference on Systems Engineering Research (CSER), March 14–16.

McCarter, Beverly Gay, and Brian E. White. 2009. "Emergence of SoS, Socio-Cognitive Aspects." Chapter 3 of *System of Systems Engineering-Principles and Applications*, edited by Mo Jamshidi. Boca Raton, FL: CRC Press, a Taylor & Francis Company.

Meadows, Donella H. 2008. *Thinking in Systems—A Primer*. Edited by Diana Wright. White River Junction, VT: Sustainability Institute, Chelsea Green Publishing.

MIT World. 2011. "Jack Is Back: A New Conversation at MIT Sloan: Never Punish Someone for Taking a Swing," http://mitworld.mit.edu/video/916

Mithaug, Dennis E. 1991. *Self-Determined Kids: Raising Satisfied and Successful Children*. New York: Lexington Books.

Moore, James F. 1996. *The Death of Competition: Leadership and Strategy in the Age of Business Ecosystems*. New York: HarperBusiness.

Morosin, Maria Simona. 2007. "Mirror Neurons Meaning and Imitation: Facts and Speculations on Language Acquisition" Studi dp Glottodidattica, 2007, 4, 90–12.

Murray, Paul. 2008. "The Power of One—Embracing and Communicating the Environmental Ethic." Systems Thinking for Contemporary Challenges Conference, Sponsored by MIT's System Design and Management Program, Massachusetts Institute of Technology, Cambridge, MA, October 23–24.

Myerson, Debra, Karl E. Weick, and Roderick M. Kramer. 2006. "Swift Trust and Temporary Groups." *Organizational Trust—A Reader*, edited by R. M. Kramer, 415–440. New York: Oxford University Press.

Nemiro, Jill, Michael M. Beyerlein, Lori Bradley, and Susan Beyerlein, Eds. 2008. *The Handbook of High Performance Virtual Teams: A Toolkit for Collaborating Across Boundaries.* San Francisco: John Wiley & Sons.

New Scientist. 2008. "Special Report: How Our Economy Is Killing the Earth." *New Scientist,* October 16. http://www.newscientist.com/article/mg20026786.000-special-report-how-our-economy-is-killing-the-earth.html?full=true

New York Times. 2010. "A Surreptitious Broadcast and a Fatal Leap." *New York Times,* September 30, http://cityroom.blogs.nytimes.com/2010/09/30/a-surreptitious-broadcast-and-a-fatal-leap/?scp=2&sq=rutgers%20student%20suicide&st=cse

Nicholls, Henry. 2011. "Quantum Evolution," *New Scientist* 28–31.

Norman, Douglas O., and Michael L. Kuras. 2004. "Chapter α, Engineering Complex Systems." The MITRE Corporation. January. http://www.mitre.org/work/tech_papers/tech_papers_04/norman_engineering/index.html

Oakley, Ed, and Doug Krug. 2006. *Leadership Made Simple: Practical Solutions to Your Greatest Management Challenges.* Greenwood Village, CO: Enlightened Leadership Publications.

O'Connell, Kevin. 2008. "The Role of Myth in Project Management." Agile Product & Project Management, Agile Advisor, Cutter Consortium, June 26, http://www.cutter.com/content/project/fulltext/advisor/2008/apm080626.html

O'Neill, Mary Beth A. 2007. *Executive Coaching with Backbone and Heart: A Systems Approach to Engaging Leaders with Their Challenges.* San Francisco: Jossey-Bass.

Ozinga, James R. 1999. *Altruism.* Westport, CT: Praeger.

Page, Scott E. 2007. *The Difference—How the Power of Diversity Creates Better Groups, Firms, Schools, and Societies.* Princeton: Princeton University Press.

Pavlou, Paul A., Yao-Hua Tan, and David Gefen. 2003. "The Transitional Role of Institutional Trust in Online Interorganizational Relationships." Proceedings of the 36th Hawaii International Conference on System Sciences.

Pfeifer, Stuart, and Ronald D. White. 2011. "FBI Raids Solar Panel Firm Solyndra after Bankruptcy Filing." *Los Angeles Times,* September.

Phillips, Helen. 2006. "Bad Habits ... That Could Help You Get Ahead." *New Scientist,* March 24.

Philips, Helen. 2006. *Instant Expert: The Human Brain.*
NewScientist.com news service, September 4.

Pierce, Eugene, and Sean W. Hansen. 2008. "Leadership, Trust, and Effectiveness in Virtual Teams." Twenty-Ninth International Conference on Information Systems, Paris.

Pink, Daniel H. 2005. *A Whole New Mind—Why Right-Brainers Will Rule the Future.* New York: Riverhead Books.

Pinker, Steven. 2009. *How the Mind Works.* New York: W. W. Norton & Company.

Plexus Institute. 2008. "Tech Savvy New Workers." *Thursday Complexity Post* (blog). Plexus Institute, Bordentown, NY, March 13. Contact: info@plexusinstitute.org

Ramachandran, Vilavanur S., and Lindsay M. Oberman. 2006. "Broken Mirrors: A Theory of Autism, Studies of the Mirror Neuron System May Reveal Clues to the Causes of Autism and Help Researchers Develop New Ways to Diagnose and Treat the Disorder." Special Section: Neuroscience. *Scientific American* (November):63–69.

Rebovich, George Jr., and Brian E. White, eds. 2011. *Enterprise Systems Engineering: Advances in the Theory and Practice.* Boca Raton, FL: CRC Press, Taylor & Francis Group.

Rizzo, Albert ("Skip"). 2008. "Clinical Virtual Reality for Mental Disorders and Rehabilitation." 8th Understanding Complex Systems Symposium. University of Illinois, Champaign-Urbana, May 13.

Rizzolatti, Giacomo, Leonardo Fogassi, and Vittorio Gallese. 2006. "Mirrors in the Mind: A Special Class of Brain Cells Reflects the Outside World, Revealing a New Avenue for Human Understanding, Connecting and Learning." Special Section: Neuroscience. *Scientific American* (November):54–61.

Robertson, Douglas S. 2003. *Phase Change: The Computer Revolution in Science and Mathematics.* New York: Oxford University Press.

Sanchez, Ron 2005. *Knowledge Management and Organizational Learning: Fundamental Concepts for Theory and Practice.* Lund, Sweden: Lund Institute for Economic Research Working Paper Series.

Sander, Todd. 2008. "Government 2.0: Building Communities with Web 2.0 and Social Networking," Digital Communities, http://www.digitalcommunities.com

Schmemann, Serge. 2006. "When the Wall Came Down—The Berlin Wall and the Fall of Soviet Communism." *New York Times*, May.

Schwarz, Roger M., Roger M. Schwarz (ed.) 2005. *The Skilled Facilitator Fieldbook*. San Francisco, CA: Jossey-Bass.

ScienceDaily. 2005. "Establishing Trust Online Is Critical for Online Communication Say NJIT [New Jersey Institute of Technology] Experts." *ScienceDaily*, June 2. http://www.sciencedaily.com/releases/2005/06/050602095433.htm

Second Life. 2008. 3-D virtual reality website homepage, http://secondlife.com/

Senge, Peter. 1990. *The Fifth Discipline: The Art and Practice of the Learning Organization*. London: Century Business.

Senge, Peter. 1990. *The Fifth Discipline. The Art and Practice of the Learning Organization*. New York: Doubleday.

Senge, Peter M. 2008. "Building Sustainable Organizations and Value Chains: What Is 'Systems Thinking' and Why Does it Matter?" Systems Thinking for Contemporary Challenges Conference, Sponsored by MIT's System Design and Management Program, Massachusetts Institute of Technology, Cambridge, MA, October 23–24.

Senge, Peter M., C. Otto Scharmer, Joseph Jaworski, and Betty Sue Flowers. 2004. "Awakening Faith in an Alternative Future—A Consideration of Presence: Human Purpose and the Field of the Future." *The SoL Journal on Knowledge, Learning, and Change*, 5(7):1–11. http://www.solonline.org/repository/download/Refl5-7.pdf?item_id=8805929

Singal, Jesse. 2008. "How to Fight a Rumor—Stopping Rumors Means Understanding Not Why They're Ugly, but Why They're Necessary." *The Boston Globe*, October 12. http://www.boston.com/bostonglobe/ideas/articles/2008/10/12/how_to_fight_a_rumor/?

Smith, Kenwyn K., and David N. Berg. 1987. *Paradoxes of Group Life: Understanding Conflict, Paralysis, and Movement in Group Dynamics*, Organization Sciences Series. San Francisco: Jossey-Bass.

Smits, Hubert. 2007. "The Impact of Scaling on Planning Activities in an Agile Software Development Context." Rally Software Development. *Proceedings of the 40th Annual Hawaii International Conference on System Sciences* (HICSS '07), January 3–6.

Snowden, David J. 2000. "Cynefin: A Sense of Time and Space, the Social Ecology of Knowledge Management." In *Knowledge Horizons: The Present and the Promise of Knowledge Management,* edited by C. Despres and D. Chauvel. Waltham, MA: Butterworth-Heinemann.

Snowden, David J., and Mary E. Boone. 2007. "A Leader's Framework for Decision Making: Wise Executives Tailor Their Approach to Fit the Complexity of the Circumstances They Face." *Harvard Business Review* 85: 68, http://www.hbrreprints.org

Stevens, Renee. 2011. *Engineering Mega-Systems—The Challenge of Systems Engineering in the Information Age.* Boca Raton: CRC Press.

Stout, Martha. 2005. *The Sociopath Next Door.* New York: Broadway Books.

Schwarz, Roger M., Roger M. Schwarz (ed). 2005. *The Skilled Facilitator Fieldbook.* San Francisco, CA: Jossey-Bass.

Surowiecki, James. 2004. *The Wisdom of Crowds: Why the Many Are Smarter Than the Few and How Collective Wisdom Shapes Business, Economies, Societies and Nations.* New York: Doubleday.

Tainter, Joseph A. 1988. *The Collapse of Complex Societies.* Cambridge, UK: Cambridge University Press.

Taleb, Nasim Nichols. 2007. *The Black Swan—Impact of the Highly Improbable.* New York: Random House.

Thomson, Helen. 2008. "Swapping Your Body Becomes a Virtual Reality." *New Scientist,* December 2. http://newscientist.com/article/dn16180-swapping-your-body-becomes-a-virtual-reality

Traut, Terence. 2008. "Characteristics of High Performance Teams," Business Resources Center—Powerful Strategies for Business Success, International Cyber Business Services, Inc., http://www.icbs.com/kb/business/kb_high-performance-teams.htm

Trevino, Linda Klebe. 1986. Ethical Decision Making in Organizations: A Person-Situation Interactionist Model. *The Academy of Management Review* 11(3):601–617. http://worldroom.tamu.edu/Workshops/CommOfRespect07/MoralDilemmas/Ethical%20Decision%20Making%20in%20Organizations.pdf

Triple Creek Associates. 2004. *Mentee Guide: Self-Paced Workbook.* Greenwood Village, CO: Triple Creek Associates, http://www.3creekmentoring.com/Mentoring_Public/Documents/Mentee_Resource.pdf

Vince, Russ. 2002. "The Impact of Emotion on Organizational Learning." *Human Resource Development International* 5:73–85.

Watson, Julie. 2008. "Researchers Re-Create Pre-Columbian Sounds—Noisemakers Made of Natural Materials Were Integral Part of Life," Technology & Science—Science, MSNBC, June 29. http://www.msnbc.msn.com/id/25391041/

Webster, Andrew. 2008. "Serious Games: Ars Looks at Games That Tackle the Big Issues," October 8, Ars Technica, http://arstechnica.com/articles/culture/serious-games-issues.ars

White, Brian E. 2006. "Enterprise Opportunity and Risk." INCOSE Symposium, Orlando, FL, July 9–13.

White, B. E. 2007. "On Interpreting Scale (or View) and Emergence in Complex Systems Engineering." First Annual IEEE Systems Conference, Honolulu, HI, April 9–12.

White, Brian E. 2008. "On Complex Adaptive Systems Engineering (CASE)." 8th Understanding Complex Systems Symposium, University of Illinois at Champaign–Urbana, May 12–15.

White, Brian E. 2008. Complex Adaptive Systems Engineering. MITRE Public Release Case No. 08-1459. 8th Understanding Complex Systems Symposium. University of Illinois. http://www.howwhy.com/UCS 2008/schedule html

White, Brian E., and Beverly Gay McCarter. 2009. "Emergence of SoS, Sociocognitive Aspects." Chap. 3 of *Systems of Systems: Principles and Applications*. Boca Raton, FL: CRC Press, Taylor & Francis Group.

White, B. E. 2011a. "Enterprise Opportunity and Risk." In *Enterprise Systems Engineering—Advances in the Theory and Practice*, edited by George Rebovich, Jr., and Brian E. White, 161–180. Boca Raton, FL: CRC Press.

White, Brian E. 2011b. "Managing Uncertainty in Dating and Other Complex Systems." Conference on Systems Engineering Research (CSER), Redondo Beach, California, April 15–16.

White, Brian E. 2011. On Principles of Complex Systems Engineering—Complex Systems Made Simple Tutorial. INCOSE Symposium, Denver, CO.

Wikipedia. 2011a. "Group Dynamics," http://en.wikipedia.org/wiki/Group_dynamics

Wikipedia. 2011b. "WikiLeaks," http://en.wikipedia.org/wiki/WikiLeaks

Wikipedia. 2011a. "Chaordic," http://en.wikipedia.org/wiki/Chaordic.

Wikipedia. 2011b. "Edge of chaos," http://en.wikipedia.org/wiki/
Edge_of_chaos.

Wikipedia. 2010. "Equifinality," http://en.wikipedia.org/wiki/
Equifinality.

Wikipedia. 2008a. "Guild," http://en.wikipedia.org/wiki/Guild.

Wikipedia. 2008b. "Power Law," http://en.wikipedia.org/wiki/
Power_law.

Woolley, Anita W., Christopher F. Chabris, Alex Pentland,
Nada Hashi, and Thomas W. Malone. 2010. "Evidence
for a Collective Intelligence Factor in the Performance of
Human Groups." *Science* 330(6004): 686–688, doi:10.1126/
science.1193147

Yee, Nick, and Jeremy N. Bailenson. 2007. "The Proteus Effect:
The Effect of Transformed Self-Representation on Behavior."
Human Communication Research 33: 271–290, http://www.
citeulike.org/user/irinas/article/1377802.

Yee, Nick, Jason Ellis, and Nicholas Duchenaut. 2009. "The Tyranny
of Embodiment." *Artifact* 2: 1–6.

Yee, Nick, Jeremy N. Bailenson, Mark Urbanek, Francis Chang, and
Dan Merget. 2007. "The Unbearable Likeness of Being Digital:
The Persistence of Nonverbal Social Norms in Online Virtual
Environments." *CyberPsychology & Behavior* 10(1): 115–121,
doi:10.1089/cpb.2006.9984, http://www.liebertonline.com/doi/
abs/10.1089/cpb.2006.9984 and http://www.ncbi.nlm.nih.gov/
pubmed/17305457

Zaki, Jamil. 2009. "The Altruism Instinct—An Antidote to the
Tragedy of the Commons." *Psychology Today*, November
23, 2009. http://www.psychologytoday.com/blog/
your-brain-us/200911/the-altruism-instinct

Zeland, Vadim. 2008. *Reality Transurfing, Volume 1: The Space of
Variations.* Winchester, UK: O Books.

Bibliography

Dirks, Kurt T., and Donald L. Ferrin. 2002. "Trust in Leadership: Meta-Analytic Findings and Implications for Research and Practice." *Journal of Applied Psychology* 87:611–28.

Ilies, R., M. W. Gerhardt, and H. Le. 2004. "Individual Differences in Leadership Emergence: Integrating Meta-Analytic Findings and Behavioral Genetics Estimates." *International Journal of Selection and Assessment* 12:207–19.

Illies, J. J., and R. Reciter-Palmon. 2002. "Destructive Leader Behavior: The Role of Personal Values." Poster presented at the 16th meeting of the Society for Industrial and Organizational Psychology, Toronto, Ontario, Canada.

Nordvik, H., and H. Brovold. 1998. "Personality Traits in Leadership Task." *Scandinavian Journal of Psychology* 39:61–64.

Van Katwyk, P. T., and P. E. Spector. 2002. "Development of an Experience Measure: The Leadership Experience Inventory (LEI)." Poster presented at the 16th meeting of the Society for Industrial and Organizational Psychology, Toronto, Ontario, Canada.

Index

286 ▪ *Index*